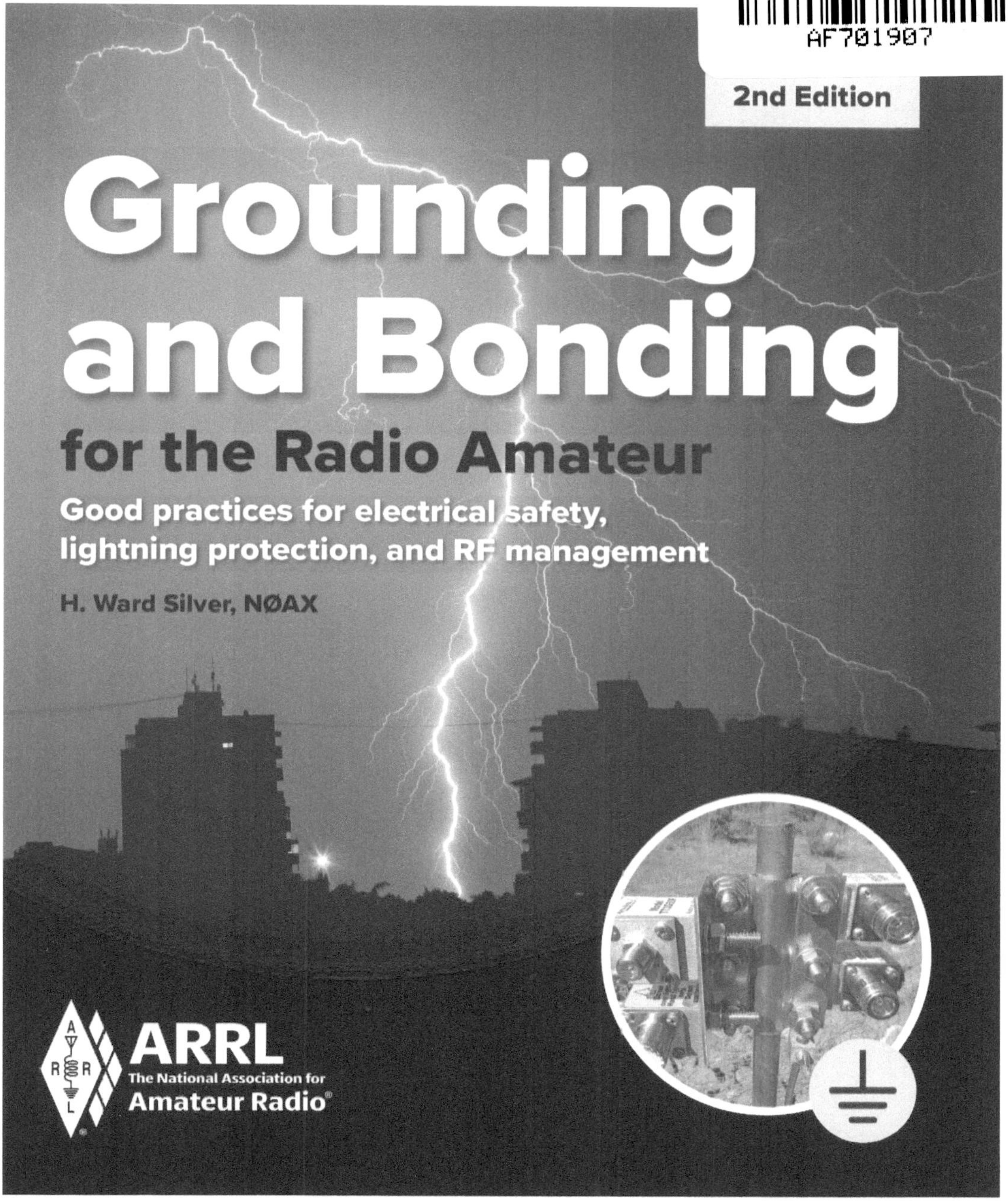

Production

Michelle Bloom, WB1ENT • Jodi Morin, KA1JPA • David F. Pingree, N1NAS • Maty Weinberg, KB1EIB

Printed in USA

ISBN: 978-1-62595-149-6

Second Edition
Third Printing

We strive to produce books without errors. Sometimes mistakes do occur, however. When we become aware of problems in our books (other than obvious typographical errors), we post corrections on the ARRL website. If you think you have found an error, please check **www.arrl.org/notes** for corrections. If you don't find a correction there, please let us know by sending email to **pubsfdbk@arrl.org**.

CONTENTS

PREFACE

The response to the first edition of this book was exceptional. Clearly, the information needed to be collected and distilled for amateur radio station-builders, and it has been well received. Numerous presentations on the subject have been given, with lots of questions and feedback contributed by the audience and by readers of this book. I have tried to add material in response to those concerns. In addition to comments from readers and viewers of the presentations, I received plenty of helpful guidance from reviewers and other experts, allowing me to refine, improve, and extend these guidelines and recommendations. Thank you to one and all!

As stated for the first edition, this book is largely a distillation of expertise contributed by others, both professional and amateur. It is based on their thousands of hours of work to determine safe and effective ways of protecting ourselves and building our stations. The author recognizes that the "deep experts" pave the way for us and encourages the reader to follow up with the references and resources listed in the book. There is much more to this subject than could ever be included in this relatively introductory book.

Along with those who developed the know-how, the author is most appreciative of the reviewers and contributors to this book: Ron Block, NR2B; Dale Svetanoff, WA9ENA; Jim Brown, K9YC; Joel Hallas, W1ZR; Alan Applegate, KØBG; Bill Wiley, WAØBSW; Bruce Carpenter, W3YVV; DX Engineering; and the ARRL Lab staff. Any remaining errors are those of the author.

I also repeat this caution to the reader — the practices and recommendations in this book are not guaranteed to prevent damage to equipment or ensure personal safety. It is the responsibility of the station builder to comply with all applicable standards and regulations, use adequately rated and listed materials, and maintain the station and equipment properly. If you are unsure of your ability to do the job correctly, hire a professional to perform or inspect your work. Where electricity is concerned, there is no substitute for careful work, attention to detail, and personal vigilance. Switch to safety!

H. Ward Silver, NØAX
St. Charles, Missouri
May 2021

About ARRL

We're the American Radio Relay League, Inc. — better known as ARRL. We're the largest membership association for the amateur radio hobby and service in the US. For over 100 years, we have been the primary source of information about amateur radio, offering a variety of benefits and services to our members, as well as the larger amateur radio community. We publish books on amateur radio, as well as four magazines covering a variety of radio communication interests. In addition, we provide technical advice and assistance to amateur radio enthusiasts, support several education programs, and sponsor a variety of operating events.

One of the primary benefits we offer to the ham radio community is in representing the interests of amateur radio operators before federal regulatory bodies advocating for meaningful access to the radio spectrum. ARRL also serves as the international secretariat of the International Amateur Radio Union, which performs a similar role internationally, advocating for amateur radio interests before the International Telecommunication Union and the World Radiocommunication Conference.

Today, we proudly serve nearly 151,000 members, both in the US and internationally, through our national headquarters and flagship amateur radio station, W1AW, in Newington, Connecticut. Every year we welcome thousands of new licensees to our membership, and we hope you will join us. Let us be a part of your amateur radio journey. Visit www.arrl.org/join for more information.

ARRL The National Association for Amateur Radio®

225 Main Street
Newington, CT 06111-1400 USA
860-594-0200
860-594-0259
membership@arrl.org

www.arrl.org

Chapter 1

Why Is This Book Needed?

When writing this book, I thought about my own experiences starting out with a simple station and wondering why grounding seemed to be so important. My radio had a two-wire ac power cord, but my vacuum-tube voltmeter had a three-wire power cord. I didn't really think a lot about it, but I noticed that connections to a ground rod were shown frequently in *The ARRL Handbook* and books on antennas.

There was a ground terminal on the back of my transceiver. The lightning arrestor I purchased had a screw for attaching a ground wire. The books and articles had lots of drawings and pictures showing quite a variety of ways to connect equipment to the ground rod. There was rarely much of an explanation about what the ground rod was actually for, or what the consequences of not having one might be — "safety" was mentioned a lot. I shrugged my shoulders, took it on faith that running a wire of some sort from my radio and antenna to a ground rod was *Something A Ham Should Do*, and just did it. I didn't notice any benefits or drawbacks from making the connection, and when the wire came loose from the transceiver on occasion, that didn't seem to matter either.

From years of operating and building all kinds of different stations, the functions of a ground rod and the connections between the equipment enclosures became clearer to me. As an electrical engineer, I learned about shielding and bonding, why the various ground connections were necessary, and the differences between them. Over the years, I built up a set of techniques that worked reliably — at home, at ARRL Field Day, in a mobile station, in basements and on upper floors, even setting up a temporary station during my 2006 trip to Brazil as a World Radiosport Team Championship competitor.

It took me years to understand how all the various things we call "grounding" really worked. From reading online forums and magazine articles, it became clear there was a lot of confusion about what grounding was intended to accomplish and how to do it. So in the January 2015 issue of *QST*, I authored the first of a series of three "Hands-On Radio" (HOR) columns, starting with "The Myth of the RF Ground." The series explained some of the differences in ground connections and why the ground connection had different functions. Those columns generated more correspondence than any other topics during the 15 years of the HOR column. I realized I'd just scratched the surface, and that is the story of how this book got its start.

Who Is This Book For?

I am imagining the reader as someone building a station and who wants to "do grounding right." This might be your first station, or perhaps you are reconfiguring your station. If you are expanding your station, you'll need to consider the impact of adding more equipment. Maybe you are getting ready for a portable or mobile operation. Or perhaps you have moved to a location where lightning is common. And there is always the unpleasant discovery that your station has an "RF hot spot" (usually right at the microphone!) that needs to be done away with. We'll talk about all of those things in the following chapters.

If you are an experienced ham, maybe even an engineer or electrician, I hope you discover some tidbits of information that make your life easier and your station work better. And I hope you'll share some of your discoveries with me so I can add them to this book's website at **www.arrl.org/grounding-and-bonding-for-the-amateur**. Amateur radio has always valued hams teaching hams, so take the time to teach other hams about grounding and bonding, won't you?

This is Not a Cookbook

I certainly wish it were possible to give clear, step-by-step instructions for making one's station safe, protected against lightning, and free of unwanted RF hot spots and RFI. But it's not. This book is not a textbook trying to present a "theory of grounding," nor is it a cookbook list of directions for all possible station configurations. It does not replace national standards or local building codes — those will always take precedence over the guidelines presented here. You'll be expected to know or learn how to work safely with the materials and tools that will be discussed.

Every station is different in the way it's constructed and wired, so just as we have a set of tools used differently to suit the circumstances, this book is a toolbox of ideas and guidelines. They are to help you meet the various electrical and mechanical requirements and constraints unique to your station.

1.1 What Are Grounding and Bonding?

The word "grounding" — meaning a connection to the Earth — is casually applied to so many different purposes in amateur radio, it's no wonder there are many opinions and misconceptions about it. "Bonding" is a less familiar term to most amateurs. In the electrical sense, bonding simply means "to connect together" so that voltage differences between pieces of equipment are minimized. **Figure 1.1** gives a couple of examples of each type of connection commonly found in an amateur's home station.

The techniques and materials needed for those connections depend on whether we are talking about ac safety, lightning protection, or RF signals. Techniques that address only one area of concern might not work for another. Clearly, this can be confusing to amateurs just trying to build a station that operates properly and safely. This book attempts to bring together basic information and best practices for the amateur from various sources.

As much as possible, this book will use standard terms. This will enable the reader to refer to standards, guidebooks, and references with a

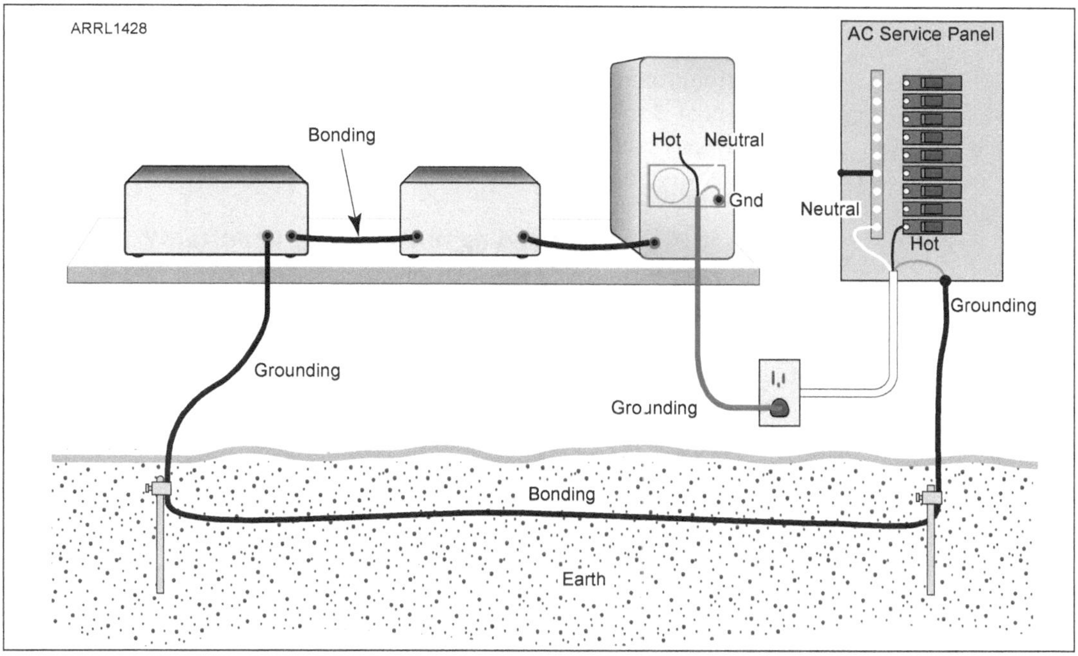

Figure 1.1 — Grounding means "to make a connection to the Earth" while bonding simply means connecting equipment or connections together to minimize the voltage between them. This figure shows a couple of examples you might find in a typical ham radio station. (These are just examples to show what bonding and grounding are.)

minimum of confusion. Amateurs often use terms inconsistently, so this book may use language differently from how we are accustomed to using it. As we progress through the book, alternate definitions or terms will be included. Regardless, I will try to use terms in a consistent way.

Grounding and bonding can be made to sound complex and difficult, but it doesn't have to be that way! Professional station designers and safety engineers have spent countless hours finding the best ways to protect equipment and operators. By following their standard practices in your station, you can get the benefit of their experiences. What is important is that you have a grasp of the fundamentals, so that you can find a way to do the right thing in the many different circumstances that amateur radio presents.

The chapters begin with the basics, including why these issues are important. After that, you'll learn the terms, the reasons for common practices, and the standards that guide them. Then we'll cover in more detail the three basic functions accomplished by grounding and bonding — ac power safety, lightning protection, and managing the RF voltages and currents in your station. When references such as websites and articles are available, they'll be listed so you can dive in even deeper if you want.

Once you are equipped with some background, examples of good practices will be provided for you to use in your station. These will be presented so you can see how to apply the principles to your own station in similar circumstances.

The goal is to help you build a better and more effective station that is protected against common hazards, and that experiences a minimum of problems from the RF signals it radiates. Let's begin.

Why Are Grounding and Bonding Important?

There are three needs we are trying to satisfy:

- *AC safety*: protect against shock hazards from ac-powered equipment by providing a safe path for current when a fault occurs in wiring or insulation.
- *Lightning protection*: minimize voltage differences between equipment during transients and surges from lightning, and dissipate the lightning's charge in the Earth, routing it away from equipment.
- *RF management*: prevent unwanted RF currents and voltages (also known as RF interference or RFI) from disrupting the normal functioning of equipment.

Satisfying these needs will also minimize potential differences between equipment at audio and power frequencies.

Each of these needs will be discussed in the following chapters, including ways of checking that your station construction will do the job. AC safety protections can, for example, be confirmed or verified by performing simple tests of protective circuit components such as GFCI circuit breakers or by an inspection, such as from a licensed electrician. Lightning protection and RF management can't really be tested short of actual experience. The time-tested approach is to follow practices that have been shown to work, and then perform an inspection to be sure the work has been done correctly.

In most instances, you will also need to comply with local building and electrical codes when installing an antenna, tower, and feed line. While these codes often incur extra expense and effort, following them helps prevent unsafe practices and lowers the risks of property damage. Your building permit and insurance may require inspection by licensed electricians, as well.

The goal of this book is to help you satisfy all of those requirements with a minimum of expense and engineering. A station constructed this way is more likely to operate safely, for longer, and with fewer disruptions and problems. As a result, you'll be able to spend more time on the air and less time troubleshooting.

Why Is "Grounding" So Complicated?

The very word — grounding — means different things depending on who you're talking to and what you're talking about. Isn't grounding just connecting equipment to the Earth? That is certainly one definition of grounding. The British use the terms "earthing" and "protective earth conductor," which are more exact references to what the connection is for. But the layer of soil and rock at the Earth's surface is not a magic zero-voltage point into which we can pour any amount of electric charge and trust that it safely disappears! The current's strength and frequency, soil characteristics, whether it is wet or dry, the length of the path to the Earth connection and through the soil — all of these affect what our equipment experiences at the "ground" connection. **Figure 1.2** shows a few of the different things that get involved when talking about "ground." As you can see, there are lots of choices!

• To the electrician, "ground" connections in your residence are a way of mitigating (or protecting) against ac shock hazards and short circuits in equipment.

• In the ac power grid, the utility uses "ground" connections along the power lines to stabilize voltage in the ac power system. This is done for lightning safety and when a fault or power system imbalance occurs.

• Ground rods outside your residence and the connections between the ground rods help minimize the damage a lightning stroke can do in your house or station.

• To an antenna system builder, the term "ground" might refer to a set

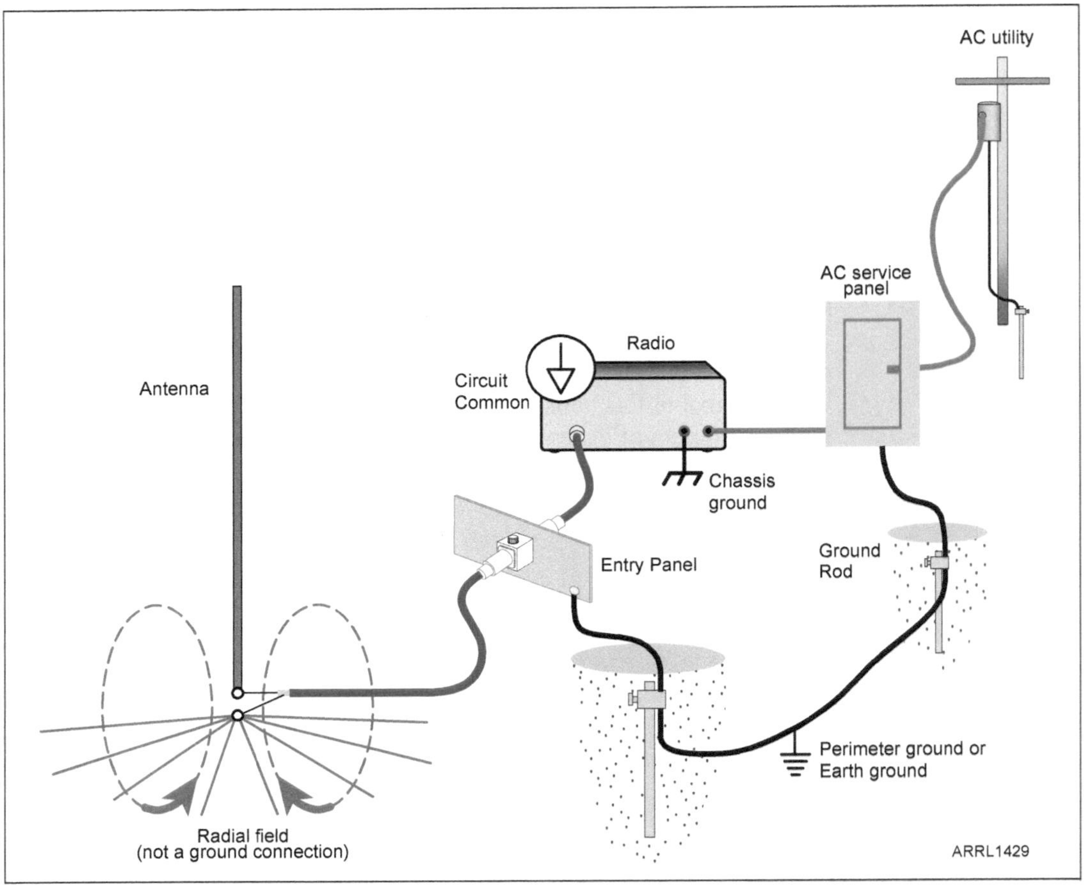

Figure 1.2 — Some of the radio station's many types of "ground" connections, not all of which are a direct connection to the Earth.

of radial wires that give RF current a path back to the feed point without flowing in the soil.

• Circuit designers and station builders refer to "ground" as a common reference voltage — Figure 1.2 shows three different schematic symbols for these connections!

The more different ways the word "ground" is used, the more different functions a "ground" connection is expected to perform, creating a pretty tall order for one piece of wire!

It is tempting to just give up, run a wire to a pipe or ground rod, and hope for the best! Most complex problems, though, can be addressed by breaking the problem down into smaller problems that are easier to deal with. The problem of simultaneously providing power safety, protecting

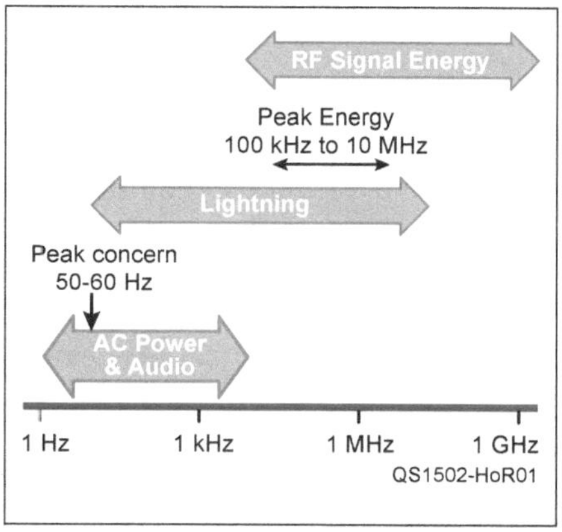

Figure 1.3 — The frequency ranges of concern to amateurs for ac power, lightning, and RF signal energy.

against lightning, and managing the effects of RF fields can be approached this way, too.

It is important to realize that grounding and bonding deal with voltages and currents spanning a very wide range of frequencies and wavelengths. **Figure 1.3** shows how big that range is. Basic electrical safety concerns involve power-grid frequencies of 50 – 60 Hz where wavelengths are 5 – 6 million meters! Audio signals span ac power frequencies to 20 kHz or more. The energy of a lightning stroke is centered around 1 MHz, spread out from dc to 10 MHz or higher. For amateurs, RF current and voltage management involve frequencies from MF into the VHF/UHF region. Amateur station wiring practices also affect tiny received signals of a few microvolts all the way to the kilo-amp and kilovolt pulses of lightning.

With such a wide variety of signals involved, no one-size-fits-all technique satisfies every need in every circumstance. By being aware of the different needs, though, you can understand each "ground" connection from the various points of view that are involved.

Seeing the Station as a System

As you read this book and try out the techniques it describes, you'll start to view your station as a big, interconnected system, and not just as individual pieces of equipment and antennas. That's just how it behaves, too, especially at audio and RF. Connections made for ac safety can have an effect on how audio and RF flow around your station, for example. By understanding each type of connection, you'll be able to build a better system that behaves the way you expect. You will also begin to see that good practices for ac safety, audio, lightning, and RF management all work together.

1.2 Materials — An Introduction

This book will provide examples and guidelines for making grounding and bonding connections. You may wonder about why certain sizes or metals are used, often heavier (and more expensive) than seems justified at first glance. In many cases, these are the sizes specified by the National Electrical Code or other industry standards. Those materials are chosen based on the hard-won experience of engineers and electricians over many years. They design systems to last for years, withstand overload and extreme events, and still protect life and property.

You may choose to use other materials for reasons of expense or convenience — that's natural for amateurs who don't have large construction budgets. It is important, though, to understand when you are varying from recommended practices, and what the possible consequences might be. Be educated, be wise, and be safe.

This is an excellent time to take a tour of your local building supply store's electrical aisles. Familiarize yourself with the various clamps, terminals, wires, compounds, and so forth. The store may have a "how-to" book on ac wiring practices for the layman — a good addition to your reference library. While you are at the store, check to see whether they have flashing in the roofing department. Speaking of libraries, you should visit your local library to become familiar with the sections on electrical practices and electronics in Dewey Decimal Classification section 621.3!

You will often find that materials purchased for electronics use are available at much lower cost at farm or rural supply stores, or anywhere they are sold in volume. If you can obtain them through a contractor or other business that buys in volume, even better. Surplus materials outlets are another possibility, if you don't mind having to hunt for what you want and buy in odd quantities or by the pound.

Another tip is to buy in quantity, particularly crimp terminals or Anderson Powerpole connectors. A bag of 100 may cost a few dollars more, but having what you need on hand when you need it makes a project much faster to complete, and saves trips to the hardware store!

Conductors at High Current

One common question is why such heavy wires are used for grounding and bonding. The reason is partly electrical and partly mechanical. These connections are required to handle worst-case current overloads without burning or melting. For example, the ground conductor of ac wiring must handle high currents for short periods of time in order to trip a circuit breaker very quickly. This makes the conductor's resistance important for ac safety.

In the case of lightning protection wiring, the short current pulse can be many kilo-amps. Obviously, even brief overloads of this magnitude would vaporize small wires! The mechanical forces between conductors carrying high currents can be enough to rip wires out of connections or break them! The force goes up with the square of the current and can become quite large.

Finally, large conductors can withstand impacts and pulling that can happen when a wire is accidentally dug up, or equipment is damaged or moved. Making a ground rod connection with a salvaged piece of #12 AWG soft copper wire might work for a while, but it could break or pull apart when hit by a shovel or mower, or even just run over by a heavy vehicle. And you might have no idea it happened! The standard #6 AWG wire may be bent, but it will probably survive the insult. Heavy conductors withstand corrosion for a long time and can also be welded to buried ground connections.

Conductors at Radio Frequencies

When we are talking about managing RF from amateur signals and the higher frequency components of lightning, the conductor's length becomes quite important. Above a few hundred hertz, the impedance of a conductor is dominated by its inductive reactance. We'll cover this subject in more detail in the appropriate chapters.

Solid copper strap, heavy wire, or soft copper tubing work well and provide plenty of surface area. Strap is often available as various sizes of scrap from metal surplus outlets. You can often get great deals for such materials at hamfests, flea markets, and garage sales, where it is often under the tables in boxes of scrap. This is a good use for pieces of ac power cable or scrap wiring. Remember that nearly all of your connections will be short, so you don't always need long lengths. Surface corrosion may not be a problem either, unless extreme, because you can brush or scrape it off to expose a clean metal surface. Keep an eye out for useful pieces and soon you'll have a good supply.

Braid versus Strap

Flat, solid strap or heavy wire are the standard for making grounding, bonding, or high-current connections at RF. Braided strap is often used for dc and low-frequency power and grounding connections — is it ever acceptable to use braided conductors for RF?

If the connection needs to flex — for example, for grounding equipment that has to be moved frequently — and it's protected from water, it is OK to use tinned flat-weave braid. Never use a braided conductor for high-frequency connections outside, or where it can get wet! Water will be drawn into the braid between the strands, where it cause the individual strands to corrode, reducing the effectiveness of the braid.

It is common practice to remove copper braid from old coaxial cable and use that for grounding — don't do it at RF! Coax braid works fine inside a cable, but when removed from its protective jacket, it rapidly begins to loosen and corrode, losing its effectiveness at RF. Just leave the jacket on the cable and create a pigtail of braid at each end so that it simply becomes a big wire.

1.3 Techniques and Tools — An Introduction

A reliable station requires reliable connections. You are probably pretty careful with your feed line and antenna connections. That attention needs to extend to the grounding and bonding connections, too! Remember that grounding and bonding connections are made for safety and need to be reliable. Use the right tools and techniques for the job. Larger tools may be available from members of your club, so you don't necessarily have to buy them.

Grounding and bonding connections use wires and cable and straps that are quite a bit bigger than the usual hook-up wire in our audio and radio electronics. You'll need to have tools that are right for the job. The hand-squeezed crimping tool in **Figure 1.4** won't do the job for large wires — unless you have a super-strong grip! The ratcheting crimper shown next to it is what you need, typically available for $30 – $40. For large wires, borrow or rent a heavy-duty crimping tool and use the right terminals for the job. Split-bolts are also suited for clamping heavy wires together and can even be used in building wire antennas.

Soldering ground connections is not a recommended practice, especially connections that may carry high-current lightning energy. The solder can melt and destroy the connection. (Some have reported success using MAPP high-temperature soldering, but this is not something for beginners.) Use clamps or crimp connections that are properly rated for this use.

Large gauge wire and sheet metal require appropriate tools to make a clean cut. Don't ruin your electronics tools by using them to cut wire or sheet too thick for them. Bolt cutters work great on those big wires and cables. Tin snips will handle sheet metal and strap. You'll also need a sturdy pair of pliers. These three tools — cutters, snips, and pliers — will get the job done most of the time. However, if a manufacturer recommends a special tool or technique, you should try to follow that recommendation both for safety and reliability.

Along with tools, it's also important to use components rated for the application and to install them correctly. Shortcuts can easily compromise safety and reliability. There are often online videos of how to do things correctly. I learned how to make the powdered "one-shot" welded connections that way! Follow the manufacturer's instructions to the letter. They want your experience with their product to be a good one. If you are confused or unsure of

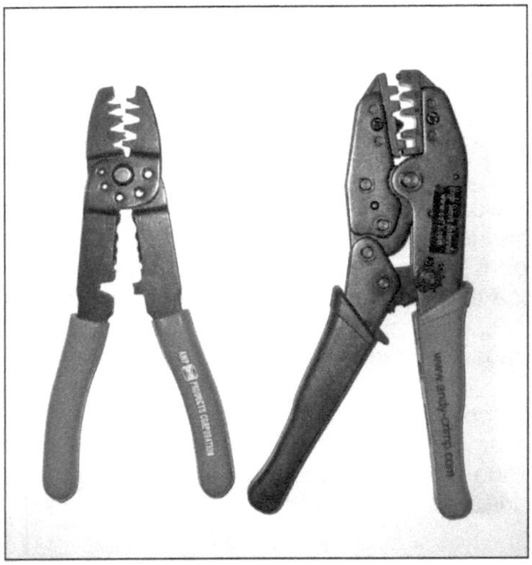

Figure 1.4 — Non-ratcheting (left) and ratcheting (right) crimping tools. The size of wires used for grounding and bonding requires the heavier tool to make a sturdy, reliable connection.

how or why something is done a certain way, an electrician or contractor might be able to explain. These connections may need to last for years, so do them right the first time!

Speaking of those welded connections (also known as "one-shots" or by the trade name of Cadweld or Uni-Shot), they are by far the most reliable method for permanent buried ground connections. Bury-rated clamps are certainly available, but a good weld lasts a lifetime. You will need to be careful, but this is a good technique to learn, and I'm glad I did.

There are also special needs for vehicle connections. Electronics in cars and trucks are subjected to wide temperature swings and are frequently exposed to water and corrosive salts. Special protective components, insulation rated for exposure to fuel and oil, and compounds that prevent corrosion are all available at your local auto parts or marine supply store. They are worth few extra dollars to keep your system solidly connected.

1.4 Things to Find Out

Before you begin to design a station layout or modify an existing layout, you should take some time to find out about local building and electrical codes that apply to your home. See the section on Important Standards and Guidelines in the next chapter. Codes are concerned with ac wiring and lightning protection, and they may be specially modified for conditions in your area. Building codes have the force of law. Being "out of code" may also affect your ability to successfully make an insurance claim or renew your insurance.

If you are a renter, your rental or lease agreement may include requirements you must meet. While hams are (or become) experts at stealthily getting on the air in such circumstances, you really don't want to do anything unsafe.

Even if you don't need a building permit or an electrical inspection, ensuring that your station "meets code" is a good way to avoid running afoul of issues you may not be aware of. The fact that you are reading this book indicates you probably aren't familiar with all of the needs for grounding and bonding. Why not take advantage of the professional expertise that went into developing those codes? At the very least, reading the code requirements gives you an idea about what issues are involved so you can make your own decisions about what to do or what not to do. A local electrician or electrical contractor may be able to provide you with basic information, and the building permit office of your town or county may also have publications that can help.

If you are planning a mobile station, read up on good wiring practices for power and bonding in vehicles. Your vehicle's dealer may have service bulletins for installing public safety or commercial radios, including preferred power connections, instructions on making connections between the radio and vehicle chassis, and even antenna placement. There may be special wiring products or connectors designed for your vehicle, as well. Here again, if you lease a vehicle, you may have to meet standards for installing radio equipment, whether it is ham gear or not. Installing electrical and radio equipment on boats has its own set of challenges — a marine supply store may have how-to guides and all the necessary parts and pieces for that environment.

When operating away from home, you need to consider basic safety requirements for using a generator, batteries, or solar power. ARRL has three books that provide guidance on power sources in the field: *Emergency Power for Radio Communications*, *Portable Operating for Amateur Radio*, and *Energy Choices for the Radio Amateur*.

This is particularly important at ARRL Field Day or a special event where several stations and antennas and power sources may all be present. Make a "dry run" in your backyard or a park to see if you have all the necessary parts and pieces to do things right.

For portable generator power, start with the two-page "Grounding Requirements for Portable Generators" from OSHA at **www.osha.gov/sites/default/files/publications/grounding_port_generator.pdf**. The manufacturer's manual should show you how to connect ground conductors and ground rods to the generator.

Chapter 2

Grounding and Bonding Basics

Before we jump into the "how to" (and "how not to") sections, let's spend some time being sure we have the basics straight. This chapter will put us on solid ground, so to speak, by explaining what a term or phrase means. I suggest that you scan this chapter before going farther — you'll sensitize yourself to important terms and the differences between them. I'll also introduce standards that cover electrical wiring practices and communications facility construction.

Along with the print edition of this book, there is a companion web page for supplemental material — **www.arrl.org/grounding-and-bonding-for-the-amateur**. Additional material, references, articles, and resources are posted on this page, and more will be added over time. The web page will include supplementary articles and downloadable versions of some of the articles referenced in the book.

2.1 Using the Right Term

If you hear or read a discussion on grounding, your first question should be, "what do you mean by ground?" There are several different types of connection called "ground," and each has a different function. Although we give each function a different name, Mother Nature does not distinguish between different connections. Current will flow wherever there is voltage and a conductive path. The following sections describe each of those functions. The goal is to help you create a single, integrated set of connections in your station that works for all three purposes: ac safety, lightning protection, and RF management.

AC or Electrical Safety Ground

Also referred to as "equipment ground," "third-wire ground," or "green-wire ground," the purpose of this connection is to prevent electrical shock and fire hazards from equipment connected to the ac power utility. The connection provides a path from the equipment back to the power distribution panel (circuit breaker box) at the ac service entry. The term "ac

Ground Symbols

There are three schematic symbols often referred to as "ground." These are shown in **Figure 2.1**. Only one (A) indicates an actual connection to the Earth. This is used for ground rods, buried bonding conductors, or other "electrodes" that make contact directly with the Earth. The symbol at (B) indicates a connection to a metallic shield or enclosure. This can be an actual metal cabinet or a small shield around a circuit. Finally, (C) shows a symbol that has many uses as a circuit common point. You may see this symbol with an A or D in the triangle, denoting that it is a common point for "analog" or "digital" circuitry. (D) shows how a coaxial connector, such as an SO-239, mounted on a metal enclosure, would be drawn.

These different symbols are often used interchangeably, but they indicate very different connections. The block diagram shows how they could be used in a typical piece of equipment. Circuit common (the closed triangle symbol) may or may not be connected to the equipment's enclosure (three-pronged symbol). The enclosure, if it is metallic, should always be connected to the ac safety ground. The safety ground is connected to the Earth (three horizontal line symbol) with a ground rod or other type of electrode. It should be clear that the symbols should only be used for their intended purpose.

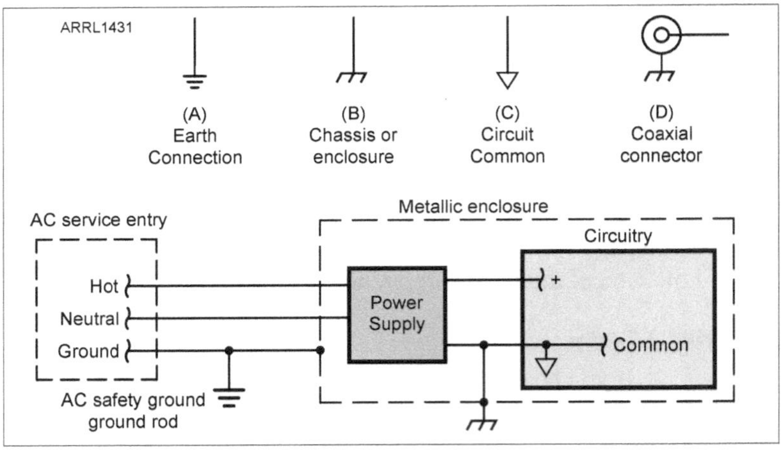

Figure 2.1 — The schematic symbols for an earth connection (A), a connection to a metallic enclosure (B), and a circuit common (C). Coaxial connectors mounted on an enclosure are drawn as shown in (D). All three of the symbols have different meanings as the typical block diagram shows.

safety ground" will be used to avoid confusion with ground connections made for other purposes.

This connection has two primary functions. First, the connection conducts enough current to trigger or trip the circuit's over-current protection if a short circuit occurs from the ac line to the equipment's exposed metal surface. The protective component could be a fuse or circuit breaker in the distribution panel or in the equipment itself.

The connection's other main function is to prevent shocks to a person touching the equipment from current flowing to ground through them. This current can be small, such as milliamps of leakage current through bypass capacitors connected to a metal enclosure. Or the shock currents can be large — an insulation failure that doesn't result in enough current to blow a fuse or breaker would clearly be enough to cause a shock. A miswired or defective ac circuit can also result in hazardous currents.

How Much Current is Dangerous?

AC currents above 50 μA flowing through the human body can be hazardous. At regular household voltage of 120 V, any resistance below 2.4 MΩ allows 50 μA or more to flow. Since the body typically presents several hundred kΩ or less between two points on the surface of the skin, any contact with household voltage is a significant hazard. (See the section on Shock Hazards in Chapter 3.)

The amount of risk from the current depends on where it is flowing. Unlike RF currents that flow along the surface of the skin, dc and low frequency ac currents flow throughout the body's tissues.

The most hazardous paths are those that go through the heart, such as hand-to-hand and hand-to-foot. This is why you should never work on equipment while barefoot or standing on a wet floor. Keep one hand clear, in your pocket for example, when the equipment is energized, to prevent a hand-to-hand shock.

Why not discuss voltage? The static voltage created by friction from clothing or just brushing hair can be hundreds or even thousands of volts! Yet the amount of charge that is actually discharged is pretty small, and so it has little effect on the heart or other muscles. On the other hand, even voltages as low as 24 V have been shown to produce hazardous current under some circumstances. The hazard comes from the effects of current flowing, not the voltage that causes the current to flow.

Lightning Protection Ground

The ground connections that protect you against lightning are a "lightning protection ground" or "lightning dissipation ground." These connections are also described as "lightning grounds" or a "lightning ground system."

A lightning stroke, which often consists of thousands of amps of current, can do a lot of damage even if it doesn't strike your house or equipment directly. Destructive currents can be induced in nearby conductors by the magnetic fields from the lightning's fast-changing current pulse, even if the strike is not a direct strike.

Lightning ground systems are designed to allow as much of the light-

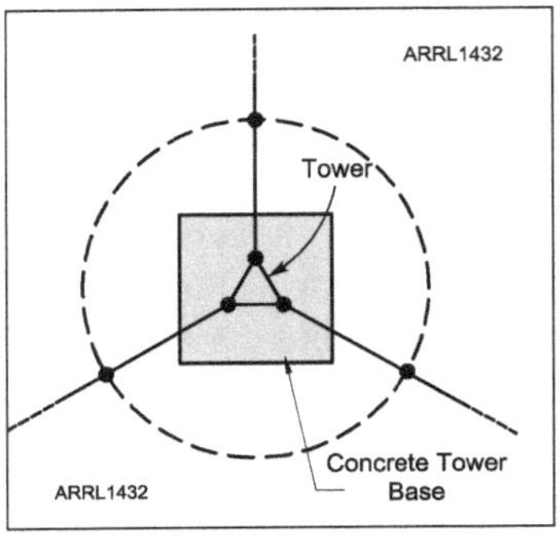

Figure 2.2 — A lightning protection ground system designed for a tower base. Multiple ground rods are connected to the base of the tower and bonded together in a ground ring (dashed circle), creating a path for charge from a lightning strike to spread out in the soil. Connections from the tower to the bonding conductor are typically extended as buried radial wires.

ning's charge as possible to spread out and dissipate in the Earth, away from structures and equipment. The larger the volume into which the charge spreads, the less likely it is that damaging voltages and currents will be present in your home and station. **Figure 2.2** shows the typical use of ground rods at the base of a tower to allow charge from a strike to spread out into the soil. Large, low-resistance ground conductors are required in a lightning ground system to withstand this large current surge without being destroyed.

A properly installed lightning ground system also helps minimize the voltage that appears between pieces of equipment during the stroke. These voltages create the arcs or transients that damage equipment. The voltage can be created directly by current from the lightning stroke flowing through the inductance of a conductor. This is one reason why short, low-inductance conductors must be used.

RF Ground and Ground Planes

The idea of an "RF ground" is really a myth. No set of wires or even the Earth itself can maintain a constant voltage across a range of frequencies above a few kHz. That term creates unrealistic expectations. What we will use instead is the idea of an "RF ground plane." (This is not the same thing as a "ground plane" used by an antenna as a conductive surface for feed point return current.)

The electrical length of connections between equipment can be a significant fraction of a wavelength on the higher HF bands and VHF/UHF bands. This allows RF voltages to be developed, causing RF current to flow between pieces of equipment. The result can be an annoying "RF burn" or erratic equipment behavior.

By connecting the equipment enclosures to a common ground or reference plane, the equipment can be kept at the same voltage as much as possible — whatever that voltage is. The plane can be a sheet of metal or simply a piece of strap or pipe to which equipment enclosures are attached. Connecting equipment enclosures together directly works well, too. The overall idea is the same — minimize voltage differences with short, low-impedance connections. In this book, we'll use the term "RF ground plane" to mean the common connection between pieces of equipment that keeps them at the same RF voltage.

Bonding

Bonding sounds like an expensive or difficult task, but all it means is connecting things together so that the voltage between them is minimized.

Building codes define bonding as a mechanically robust, low-impedance connection capable of carrying the expected fault current. The challenge is to be sure the connections are electrically and mechanically suitable. When discussing ac safety and lightning, whatever is used for bonding conductors has to be heavy enough to handle the substantial currents. Lightning, RF, and audio have a very wide frequency range, so the impedance of the conductor must be low, as well.

For an indoor ac safety ground, the equipment ground conductor must be at least as large as the energized ("hot") conductors in the circuit supplying power. For creating an RF ground plane in your station or connecting to it, strap or wire such as #12 or #14 AWG can be used. For a lightning dissipation ground, usually outdoors and buried, connections may require #6 or #2 AWG copper wire at a minimum. Use the proper materials, terminals, and clamps to secure the connections and you will be in good shape. We'll see examples of terminals, clamps, and other materials used for this purpose later in this book.

Single-Point Ground

The term "single-point ground" refers to a shared connection point also connected to the Earth. The single-point ground is assumed to be small enough and conductive enough that it can be treated as a zero-ohm connection with the same voltage everywhere on it, whether it is a ground rod, a metal strap, or a sheet of metal.

"Electrically small" depends on what frequencies are involved, though. At ac power frequencies, where the wavelength is miles long, the dimensions of the connection are insignificant and resistance is the main concern. For lightning, where the primary spectrum is approximately 100 kHz to 10 MHz, the inductance of the shared connection is the most important thing. And for controlling RF voltage in your station, both inductance and size matter. What the electrician and the amateur consider single-point grounds can be quite different!

A related term, "star ground," refers to using a single point of connection for circuits that share a common ground. Electronic circuits might use a star ground on a PC board to avoid introducing noise into a sensitive sub-circuit. **Figure 2.3** shows a star ground for several pieces of equipment. The star

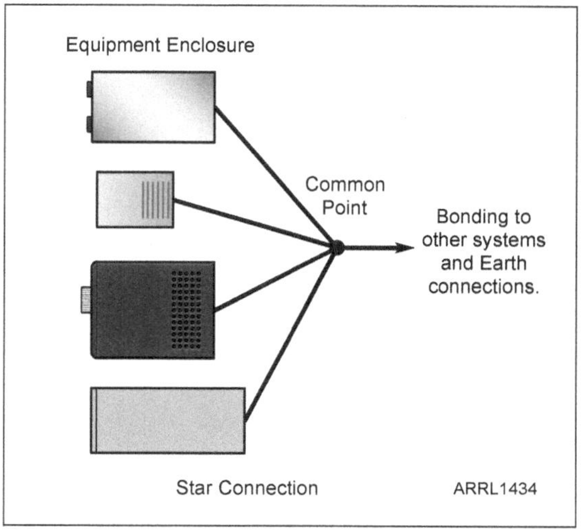

Figure 2.3 — A star ground connection used for several pieces of equipment sharing a common ac safety ground.

ground is effective for ac safety and, if the bonding conductors are short and direct, also effective for lightning protection. Star grounds are a good technique for grounding a tower and associated equipment. Using a star ground for RF management is somewhat unpredictable due to the higher frequencies involved, so other techniques will be discussed.

Single-Point Ground Panel

A "single-point ground panel" or SPGP (see **Figure 2.4)** is an externally grounded metallic panel, typically installed near the radio equipment along the natural flow of the cabling from where it enters the room or building to where it connects to the equipment. If installed at an exterior

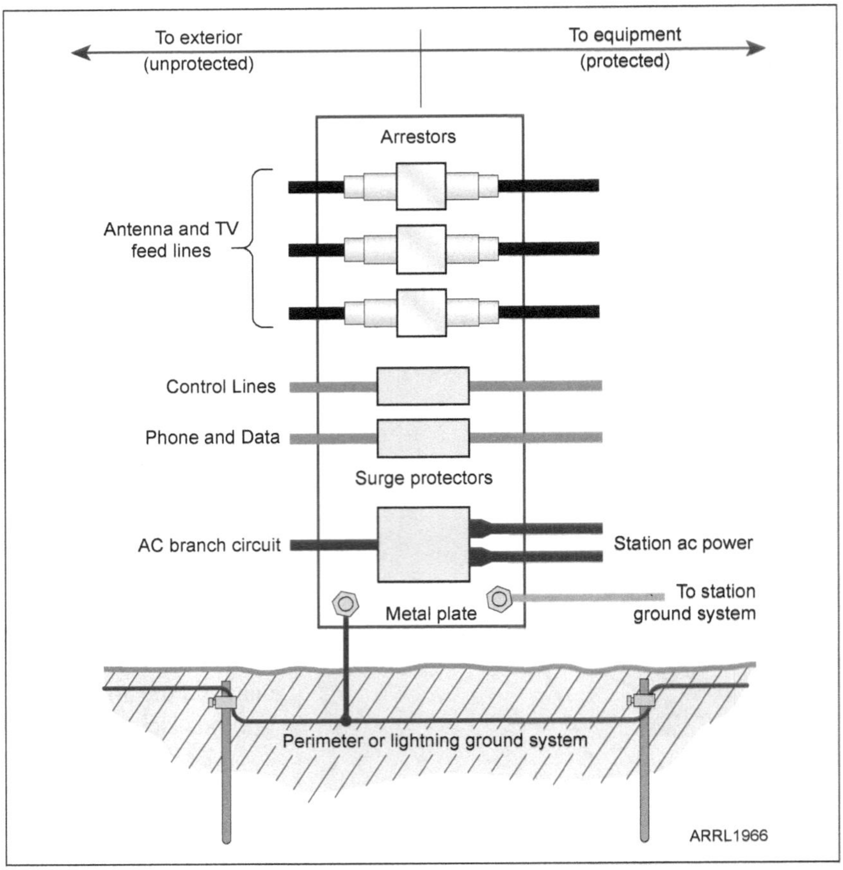

Figure 2.4 —A single-point ground panel (SPGP). Arrestors and surge protectors for feed lines, control lines, telephone and data services, and ac power are mounted on the panel, which is connected to the station ground system and the external perimeter or lightning ground system. The SPGP is discussed in more detail in Chapter 4.

wall, it may also serve as an "entry panel" or "entrance panel." The SPGP is bonded to a ground system as described in the following chapters. Surge protectors for the coax cables, rotator control, and ac-power are mounted on the SPGP. Cables external to the SPGP that may be exposed to lightning energy are called *unprotected*. Cables between the SPGP arrestors or surge protectors and the equipment are called *protected*.

The function of the surge protectors and arrestors, when over-voltaged by strike energy, is to short all the wires within each cable to the SPGP. With all the external wiring having a protector mounted on the SPGP, all cables connected to the radio equipment are shorted together so that electrically the equipment appears to be connected by a single wire. Thus, no current will flow within or between pieces of equipment. This is the secret of success! Construction and use of the SPGP is discussed in more detail in Chapter 4.

Shall, May, and Should

Readers who deal regularly with requirements or contracts understand the difference between these three words. They are very important in standards, permits, and instructions — here are typical interpretations:

- Shall — you *must* do this to comply
- May — doing this is optional, and you will have complied whether you do it or not
- Should — this is recommended, but not required to comply

It is unlikely that you'll be reading a standard at this level of detail, but if you encounter these words in a building regulation, for example, be aware of what they mean. Confusing a "shall" for a "should" can cause a lot of expensive rework!

Authority Having Jurisdiction (AHJ)

This term has a very specific legal meaning — the government agency with the authority to enforce a rule. Before beginning any changes in permanent wiring, be sure you know what the AHJ is, and what rules you

Symbols for Antennas and Towers

Throughout this book, you'll see drawings of a tower-and-beam combination used to represent the external antenna system. This refers to any outside antenna, including wire antennas, ground-mounted verticals, and roof-mounted antennas of all types. The tower represents any conductive antenna support structure, including lightweight TV masts or aluminum "push-up" poles. Where something applies to a specific type of antenna or tower, an appropriate drawing or text reference will be used.

must follow. For example, the AHJ for electrical wiring in your home is whatever agency issues building permits. (If you are a renter, your landlord may impose additional conditions but is not considered an AHJ.)

The Earth, Ground, and Zero Volts

"The Earth is not a sink into which we can pour RF current or noise and simply have it disappear." (Jim Brown, K9YC, and others) Across the wide range of frequencies, currents, and voltages encountered by amateurs, there is no way the Earth can be considered as a constant voltage, also known as an *equipotential surface*. Voltage drops from large lightning currents create hazardous voltage differences between points just feet apart in the vicinity of the stroke. At the upper end of the HF spectrum, moving even a few feet is enough to shift from a "hot spot" to a "dead spot." All sorts of voltages resulting from events outside our station can exist anywhere on what we think of as "ground."

Remember that voltage is always measured between two points. They can both be on the Earth. Or one point can be on your body and the other on the Earth. Or they can be on different pieces of equipment and *neither* be on the Earth. Especially at RF, what you might consider to be "zero" volts over *here* is unlikely to be "zero" volts over *there*. So we have to give up the notion that a connection to the Earth's surface will present a constant voltage with respect to all other points.

It's a lot easier to think about the Earth in terms of what happens to the current flowing in or on it. For ac safety, the "ground" rod provides a connection to the Earth for voltage and current transients ("spikes") on your ac wiring from either external or internal causes. For lightning, the "ground" is there to dissipate charge into the Earth as effectively as possible. For RF, treat the surface layer of the "ground" as a thick, lossy sheet of resistive and capacitive material.

2.2 Important Standards and Guidelines

A quick internet search for standards on grounding will turn up dozens of official-sounding documents from many different organizations. The two standards described below are widely used — the *National Electrical Code* (NEC, or NFPA 70) published by the National Fire Protection Association, and *Standards and Guidelines for Communication Sites* (or R56), published by Motorola. You might find copies, but they are not "how-to" guides intended for use by non-professionals. If they were, this book would not be necessary — you could just look at the standards. To give you clearer instructions and guidance, the material in this book is collected from many articles and guidelines that interpret these and other standards.

National Electrical Code (NEC)

The NEC is a set of standards that can be adopted by local building codes for electrical wiring in the home and businesses. It has been adopted as an American national standard by the American National Standards Institute (ANSI). Don't confuse the NEC with your local building code. A casual reference to "the code" might mean the NEC or your local building code. The building code always takes priority. The NEC is organized in chapters that cover a general topic and in various *articles* that cover a specific sub-topic.

There are two versions of the NEC: the code and the handbook (**Figure 2.5**). The code version is just that — the text of the standard only. The more expensive handbook goes farther and gives drawings and examples that are compliant with the code. A personal copy of the *NEC Handbook* is expensive, about $200 for new copies, so you should check out your local library or building department to find a copy you can consult when needed. Even with the handbook's explanation, you will still need some interpretation and guidance about what the code requires you to do.

The website **www.mikeholt.com** also has quite a bit of information about the NEC and wiring practices. For example, you can learn a lot about the NEC from **www.mikeholt.com/understanding-the-nec.php**.

If the NEC is adopted by municipalities as part of its building code, it becomes law as discussed in Chapter 3. While local variations are common, most building codes and the parts that most apply to hams use the regular NEC language. If you follow the NEC when constructing your station, you are almost certainly going to be in compliance with your local building rules. That doesn't mean you need to become an expert electrician. Most of the basic rules are easy to learn and apply. Take care to do things as directed and to use the right materials.

Figure 2.5 — The 2020 revision of the *National Electrical Code*. The handbook version supplements the code with additional drawings and schematics. Check your library for the latest version.

First and foremost, be safe! The NEC wasn't created just to make your life difficult. Its knowledge of how to do things right was gained by the practical experience of thousands of electricians and engineers. When in doubt, consult a local, licensed electrician or contractor or inspector to advise you on what to do.

The NEC has chapters on basic power wiring (ac and dc) that go into great detail about specific practices. You will probably not need to do much more than check out or wire up some ac outlets, use a generator or solar panel, or run new power circuits to your station. Safe practices for doing these basic jobs are illustrated in numerous "how-to" books available where you purchase electrical supplies or from a library or book seller. (See Chapter 3 for some suggestions.) In the NEC itself, you'll find these topics covered in Chapters 1 – 4.

Chapter 8 of the NEC covers communications systems, and Article 810 is specifically about radio and television systems. This article presents a lot of good information about the proper grounding of external antennas and towers for both ac safety and lightning protection. The chapter also covers antenna and feed line installation practices, including electrical and mechanical safety.

Standards and Guidelines for Communication Sites (R56)

This commercial standard is a comprehensive "how to" for designing, constructing, and maintaining communication system sites. The 2005 edition (**Figure 2.6**) is widely available as a downloadable document from websites such as **www.blm.gov/sites/blm.gov/files/Lands_ROW_Motorola_R56_2005_manual.pdf** or **www.repeater-builder.com/antenna/site-stuff/are-fifty-six-man-2005.pdf**. These communication sites primarily use rack-mounted VHF and UHF equipment with towers next to the building. Nevertheless, the standard provides useful guidelines for typical amateur stations, as well.

Figure 2.6 — Motorola's R56 *Standards and Guidelines for Communication Sites* is a comprehensive guide for building robust stations.

The primary chapters of interest to this book's readers are on External Grounding (Earthing) – Chapter 4, and Internal Grounding (Earthing) – Chapter 5. There is also a chapter on Surge Protective Devices (Chapter 7) that applies to discussions of lightning protection. You'll note that the standard uses the term "grounding (earthing)" throughout. This reinforces the one specific meaning of the term, referring to an electrical contact to the Earth that the NEC calls a ground or earth electrode.

The standard covers so many topics related to station construction that it is a good reference to have handy. Even if you don't have the ability (or the budget) to follow its recommendations exactly, it might be a good idea to read the appropriate material in the standard to get an idea of how the pros do it and see how you might come close. At the beginning in its Table 1-1, there is a comprehensive list of related safety and practice standards that you can refer to, as well.

Military Standard MIL-HDBK-419A

The primary military standard for grounding and bonding in communication systems is *Grounding, Bonding, and Shielding for Electronic Equipments and Facilities*, MIL-HDBK-419A, released in 1987. The standard is composed of two volumes that deal with grounding, bonding, and shielding theory for communications electronics equipment and facilities. Theory is covered in the first volume and applications in the second.

Volume 1 is of most interest to amateurs. It covers the principles of personnel protection, fault protection, lightning protection, interference reduction, and EMP protection. In addition, the basics of earth connections, signal grounding, electromagnetic shielding, and electrical bonding are presented. The subjects are considered from the standpoint of how they influence the design of a facility's earth electrode subsystem, designing ground reference networks for equipment and structures, shielding requirements, bonding practices, and so on. There are many useful diagrams, tables, and charts in this document. The Motorola R56 standard uses similar approaches in several areas.

The document is in the public domain and can be obtained from online sources such as **https://ntrl.ntis.gov/NTRL/dashboard/searchResults/titleDetail/ADA239565.xhtml** (Volume 1). Search for MIL-HDBK-419A.

Chapter 3

AC Power System Grounding and Bonding

The material in this chapter is based on the Electrical Safety section of *The ARRL Handbook* and the appropriate articles (sections) of the *NEC Handbook*. Each of the main points will be covered and explained in simple terms to help you get started and build your station with the appropriate safety features.

What Do I Need to Learn from This Chapter?

- What ac safety grounding is, and why it is necessary
- The basics of safety grounding for household ac power wiring
- How to connect your station's ac safety ground properly
- How to power your station safely from ac power

This chapter is only concerned with ac safety grounding, and so it is primarily concerned with helping you understand and satisfy the NEC requirements. These are what building codes in the United States use, covering all of the necessary practices for stations in a house or apartment. Where special practices are involved that are unique to radio communications equipment, guidelines from the Motorola R56 standard are referenced. See the Grounding and Bonding Basics chapter for more information about *Standards and Guidelines for Communication Sites* (R56).

But first…

3.1 Safety First — Always

While constructing an effective station with limited resources requires ingenuity at times, there is still a need to act responsibly. Safety begins with your attitude about it. Make it a habit to plan work carefully and take the time to do the job right. If you don't have the right materials or tools, get them. Don't rely on temporary measures — they often become permanent. Your project is not a race!

National Electrical Code

Exactly how does the National Electrical Code become a requirement? How is it enforced?

Cities and other political subdivisions have the responsibility to act for the public safety and welfare. To address safety and fire hazards in buildings, regulations are adopted by local laws and ordinances — usually including some form of permit and accompanying inspections. Because the technology for the development of general construction, mechanical, and electrical codes is beyond most city building departments, model codes are incorporated by reference. For electrical issues, the National Electrical Code is in effect in virtually every community. City building officials will serve as "the authority having jurisdiction" (AHJ) and interpret the provisions of the Code as they apply it to specific cases.

Building codes differ from planning or zoning regulations: Building codes are directed only at safety, fire, and health issues. Zoning regulations often are aimed at preservation of property values and aesthetics. The NEC is part of a series of reference codes published by the National Fire Protection Association, a nonprofit organization. Published codes are kept up to date regularly and are developed by a series of technical committees whose makeup represents a wide consensus of opinion. The NEC is updated every three years. It's important to know which version of the code your local jurisdiction uses, because it's not unusual to have the city require compliance to an older version of the code. Fortunately, the NEC is usually backward compatible: that is, if you're compliant to the 2020 code, you're probably also compliant to the 2017 code.

Do I have to update my electrical wiring as code requirements are updated or changed?

Generally, no. Codes are typically applied for new construction and for renovating existing structures. Room additions, for example, might not directly trigger upgrades in the existing service panel unless the panel was determined to be inadequate. However, the wiring of the new addition would be expected to meet current codes. Prudent homeowners, however, may want to add safety features for their own value. Many homeowners, for example, have added GFCI protection to bathroom and outdoor convenience outlets.

(Adapted from The ARRL Handbook, Safe Practices chapter)

You Are a *Lot* Slower Than Electricity!

We have all experienced the temptation to make a very quick "swipe" of a wire to see if it is "hot." This is not a good way to learn about how fast electricity can move. With effort, you might be able to move your finger a few feet per second, but electricity moves millions of times faster. As fleeting as the contact might be to you, there is *plenty* of time for current to flow and deliver a full-strength shock. So let's put that temptation right out of our thoughts — just don't — and be sure your kids know that, too, because there might not be a second chance.

Safety guidelines cannot possibly cover all situations, but by approaching each task with a measure of understanding about electrical safety, it should be possible to make sure your station is safe and that you can work on it safely. Don't be the one to say, "I didn't think it could happen to me."

What if something bad does happen? Having other people around is always beneficial in the event that you need immediate assistance. Take the valuable step of showing family and club members how to turn off the electrical power safely. Additionally, cardiopulmonary resuscitation (CPR) training can save lives in the event of electrical shock. Classes are offered in most communities. Take the time to plan with your family exactly what action should be taken in the event of an emergency, such as electrical shock, equipment fire, or power outage. Then practice your plan!

3.2 Shock Hazards

Electrical shocks are current — either ac or dc — flowing through the human body. The NEC and other building codes are written specifically to prevent electrical shock, to prevent electrical fires, and to provide lightning safety. It is important to follow these requirements carefully. Don't become a statistic.

There are lots of ways to get a shock, of course, and there are many good safety practices to prevent them. *The ARRL Handbook* contains information on safety concerns about high-voltage (HV) power supplies and batteries, for example. This chapter only discusses hazards caused by

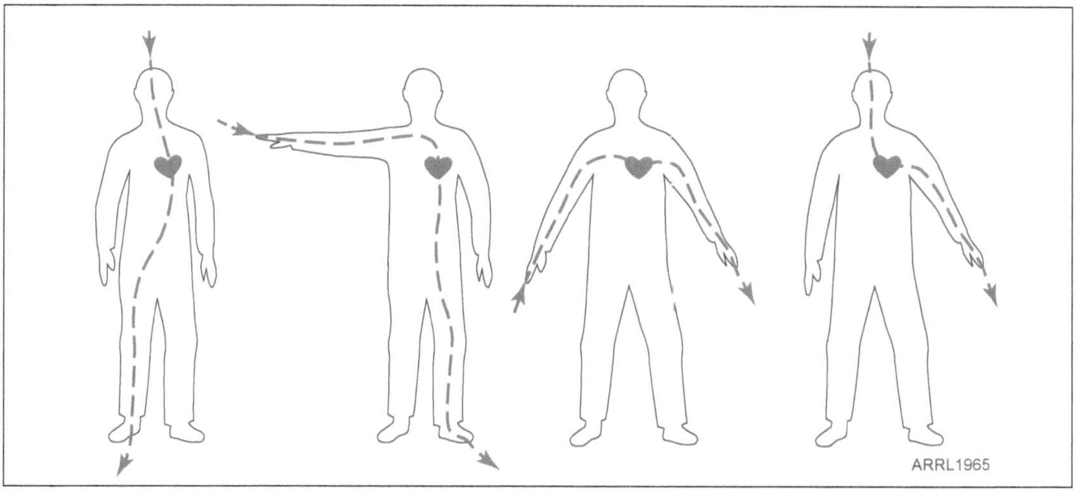

Figure 3.1 — The most dangerous paths for shock current through the body go through the heart. This can disrupt the heart's electrical rhythm. The most common shock paths are hand-to-hand and hand-to-foot.

equipment that is connected to the ac power system. Taking care of these hazards probably helps with hazards from other electrical sources, too, but we're going to focus on ac power.

Why does current flow through the body? The current path is created when two points on your skin are in contact with wires or other metal surfaces at different voltages. The most common paths are between the two hands or between a hand and the feet (**Figure 3.1**). Both paths allow current to flow through the heart, which is particularly dangerous because it can disrupt the heart's electrical rhythm.

Figure 3.2 — Keep one hand in a pocket or behind your back when testing energized equipment to avoid creating a path for shock current.

To prevent this type of shock current path, keep one hand in a pocket or behind your back when making a measurement on energized equipment (**Figure 3.2**). Similarly, never work barefoot around energized equipment — make sure your shoes are dry and have insulating soles. Don't give current a path you don't want it to take.

The previous chapter points out that any current larger than 50 μA indicates a hazard may be present, so the resistance of the body and how contact is made are important according to Ohm's Law: $I = V/R$. The sidebar "Electrical Shock Hazards" is a good summary of the various factors and effects of shock.

Most sources start measuring the effects of shock from ac at 50 V, and voltages as low as 24 V have been able to cause the sensation of shocks under certain circumstances. It's safest to assume that much above the relatively benign 12 V used by most vehicles and solid-state transceivers, there is no such thing as a "safe" level of voltage. Hams have gotten used to 12 V being the highest voltage in a piece of equipment, and it's easy to get complacent. Be extra careful around ac line voltages and any vacuum tube circuits. Vacuum tubes may operate at hundreds or even thousands of volts.

Where Can Hazards Exist?

A shock hazard is possible in any piece of equipment connected to a source of ac voltage. Insulation can fail, a wire or terminal can be pushed against a metal panel, or conducting material (even dust!) can build up between a live circuit and the enclosure. There are myriad ways for such connections, called *faults*, to be created. The current flowing as a result of a fault is called *fault current*. Fault currents in a home's ac wiring can be large enough to trip a breaker from excessive current, or they can be *leakage currents* that range from barely perceptible to painful.

Any current through the body should be considered a shock hazard, so it's best to just treat all contact with an *energized* metal surface as something to avoid. The important idea behind bonding for electrical safety is that proper grounding de-energizes these surfaces by causing power to be removed (tripping that circuit breaker, for example) or giving current an alternate path to ground that doesn't go through you.

In an amateur station, ac shock hazards are presented by any piece of equipment connected to the ac line, such as a power supply, computer equipment, antenna rotator, amplifier, battery charger, and so on. Equipment that is not ac-powered but connected to something that is can present a hazard, too. For example, an SWR bridge with a metal case that is connected to an ac-powered amplifier's output with a coaxial cable has a direct connection to the amplifier enclosure. If the amplifier enclosure becomes energized, so does the enclosure of the SWR meter! With so many interconnections in even simple amateur stations, it's best to treat all exposed metal the same — enclosures, connectors, cable shields, keys (especially those connected to vacuum tube equipment), and microphones — and ensure that it is properly bonded.

What is Mitigation?

With respect to electrical safety, *mitigation* means to make a safety hazard, or consequences of the hazard, less severe. In plainer words, mitigation means to reduce the likelihood of a hazard existing at all, and if it does exist, lessen its possible consequences.

Mitigating against a shock hazard can take several forms:

1) Eliminate the possibility of getting shocked through equipment and wiring design. For example, use a properly constructed and protected 12 V dc power supply instead of 120 V ac. This obviously gets rid of the shock

Test Instrument Ratings — Know and Heed!

Handheld meters and other instruments have a maximum voltage rating — do you know what it means? It is not the maximum voltage the meter can read. It is the maximum voltage that can be measured safely! Higher voltages can result in shock from a *flashover* — an arc from the test leads or terminals to the user. Stay within the meter's ratings to be safe. (See the accompanying photo.)

Cracked, frayed, loose, or otherwise damaged test leads and terminals can also present a hazard, even if the voltage is within the meter's rating. Inspect your meter's case and leads to be sure they are in good condition before use on hazardous voltages.

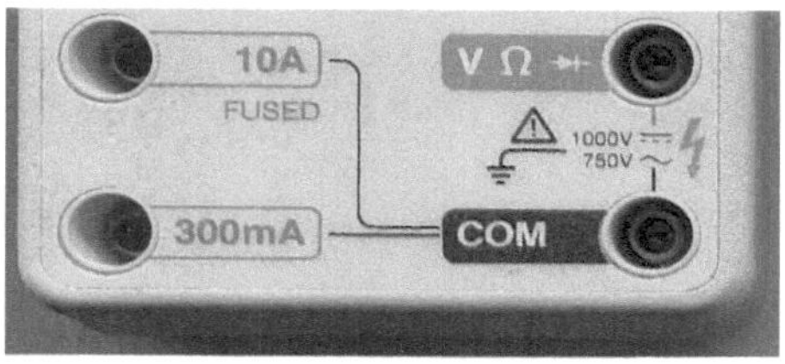

Voltage ratings are shown next to a voltmeter's probe jacks (1000 V dc and 750 V ac for this meter). Exceeding these voltages is a severe electrocution hazard.

Electrical Shock Hazards

(The following material is adapted from the website "Electrical Shock Hazards, Risks, and Other Dangers" published by Safety Protection Grid Solutions at **www.spgsamerica.com/information/shock-hazard**.)

An electrical shock to a human can be as mild as a slight tingling sensation or as severe as instant death. Ohm's Law will determine how much electrical current will flow on the grounding conductor and the human body. This electrical current flowing through the human body will create some type of shock to the human.

To better understand the amount of amperage that will affect the human body and how the body reacts to these amperage levels, we must first review the examples of body resistance of human contact points under different conditions, as shown in **Table 3.A**.

The conditions necessary to produce 1000 Ω of body resistance don't have to be as extreme as what is presented, either (sweaty skin with contact made on a gold ring). Body resistance may decrease with the application of voltage so that with constant voltage a shock may increase in severity after initial contact. What begins as a mild shock — just enough to "freeze" a victim so they can't let go — may escalate into something severe enough to kill them as their body resistance decreases and current correspondingly increases.

AC is more dangerous than dc, and 60-cycle current is more dangerous than high-frequency current. Skin resistance decreases when the skin is wet or when the skin area in contact with a voltage source increases. It also decreases rapidly with continued exposure to electric current.

The real measure of a shock's intensity lies in the amount of current forced through the body and not the voltage. Any electrical device used on a house wiring circuit can, under certain conditions, transmit a fatal current. The resistance of the human body varies so widely it is impossible to state that one voltage is "dangerous" and another is "safe."

The path through the body has much to do with the shock danger. A current passing from finger to elbow through the arm may produce only a painful shock, but that same current passing from hand to hand or from hand to foot may well be fatal. Therefore, the practice of using only one hand (keeping one hand behind your back) while working on high-voltage circuits, and of standing or sitting on an insulating material is a good safety habit. (See Figure 3.2.)

Some people are highly sensitive to current, experiencing involuntary muscle contraction with shocks from static electricity. Others can draw large sparks from discharging static electricity and hardly feel it, much less experience a muscle spasm. Despite these differences, approximate guidelines have been developed through tests which indicate very little current being necessary to manifest harmful effects.

The Physiological Effect of Electric Shock

Electric current damages the body in three different ways: (1) it harms or interferes with proper functioning of the nervous system and heart; (2) it subjects the body to intense heat; and (3) it causes the muscles to contract. **Table 3.B** summarizes the effects of current at different levels. Note that none of the currents listed below would trip a 15 A breaker or fuse.

Table 3.A
Typical Body Resistance

Type of Contact	*Dry*	*Wet*
Wire touched by finger	40,000 – 1,000,000 Ω	4000 – 15,000 Ω
Wire held by hand	15,000 – 50,000 Ω	3000 – 5000 Ω
Metal pliers held by hand	5000 – 10,000 Ω	1000 – 3000 Ω
Contact with palm of hand	3000 – 8000 Ω	1000 – 2000 Ω
1.5 inch metal pipe grasped by one hand	1000 – 3000 Ω	500 – 1500 Ω
1.5 inch metal pipe grasped by two hands	500 – 1500 Ω	250 – 750 Ω
Hand immersed in conductive liquid		200 – 500 Ω
Foot immersed in conductive liquid		100 – 300 Ω

Table 3.B
Effects of Current at Different Levels

AC (mA at 60 Hz)	*DC (mA)*	*Effect*	*Body Resistance Required to Create Current at 120 V ac*	*Body Resistance Required to Create Current at 50 V dc*
0.3 (Women)	0.6	Slight sensation felt at hand(s)	400,000 Ω	83,333 Ω
0.4 (Men)	1	Slight sensation felt at hand(s)	300,000 Ω	125,000 Ω
0.7 (Women)	3.5	Threshold of perception	171,428 Ω	14,285 Ω
1.1 (Men)	5.2	Threshold of perception	109,090 Ω	9615 Ω
6 (Women)	42	Painful, but voluntary muscle control maintained	20,000 Ω	1190 Ω
9 (Men)	62	Painful, but voluntary muscle control maintained	13,333 Ω	806 Ω
10.5 (Women)	51	Painful, to let go of wires	11,428 Ω	980 Ω
16 (Men)	76	Painful, to let go of wires	7500 Ω	657 Ω
15 (Women)	60	Severe pain difficulty breathing	8000 Ω	833 Ω
23 (Men)	90	Severe pain difficulty breathing	5217 Ω	555 Ω
25 (Women)		Painful shock, muscular control is lost	4800 Ω	
30 (Men)		Painful shock, muscular control is lost	4000 Ω	
20-75		This shock is more serious. You'll receive a painful jolt and muscle control will be lost resulting in the inability to let go of something you may have grabbed that is shocking you	1600 Ω	
75 – 100		As the current approaches 100 milliamperes, ventricular fibrillation of the heart occurs and damage is done	1200 Ω	
100 (Women)	500	Possible heart fibrillation after 3 seconds	1200 Ω	100 Ω
100 (Men)	500	Possible heart fibrillation after 3 seconds	1200 Ω	100 Ω
100 – 200		Ventricular fibrillation occurs and death can occur if medical attention is not administered quickly	600 Ω	
>200		Severe burns and severe muscle contractions occur. Your heart can stop during a shock because the chest muscles put pressure on the heart. Internal organs can be damaged at this stage and in you survive, a painful recovery can be expected. What may surprise you about this level of shock is that through this clamping effect on the heart, ventricular fibrillation is avoided and the chances of a person's survival is good if the victim is removed from the electrical circuit	600 Ω	
1000 – 4300		Ventricular fibrillation. (The rhythmic pumping action of the heart ceases.) Muscular contraction and nerve damage occur. Death is most likely.	27.9 Ω	
6000		Sustained ventricular contraction followed by normal heart rhythm. (Defibrillation). Temporary respiratory paralysis and possibly burns.	20 Ω	
10,000		Cardiac arrest, severe burns, and probable death	0.012 Ω	

Concrete As a Conductor

Regular concrete is not a very good conductor, but it is still a better conductor than many types of soil. If the concrete is damp or in contact with damp soil, its conductivity will increase. Steel rebar in a concrete slab or wall will also increase conductivity. (See the Chapter 6 section on Ufer grounds.) There are also additives that make concrete quite conductive for various special purposes. It is important to realize that concrete, even if it is dry on the surface, can be a good enough conductor to present a shock hazard if you are in contact with it and a piece of energized equipment. You should never work on live equipment while barefoot on concrete surfaces or touching concrete.

hazard in that piece of equipment, but is not always possible. Where does the 12 V dc come from? Higher current at 12 V might have its own associated fire hazard. It's hard to eliminate all hazards completely.

2) Prevent encountering an electrical shock hazard through construction practices. Since hazardous voltages are part of using electrical power, prevent mechanical or electrical failures from exposing a person to them. Connect metal enclosures to the power system's safety ground so the current doesn't flow through the person. (Double-insulated power tools are an example of Class II equipment — those having no exposed metal parts, or exposed metal parts that are double-insulated from mains power.)

3) Reduce the effect of the hazard. If a person does encounter the hazard, try to limit the magnitude of the shock by keeping the current below hazardous level. Alternatively, detect the shock current and de-energize the circuit it's coming from as quickly as possible, such as by using a Ground Fault Circuit Interrupter (GFCI).

Eliminate, protect, reduce — words to remember as you assemble your station! Ask yourself, "What would happen if…" and take the appropriate steps.

3.3 Power System Grounding and Bonding

We won't try to make this chapter an instruction manual for ac home wiring. That is well-covered in a variety of books and websites. What this section discusses is the requirements for ac safety grounding in your station. Your local building codes set the standard for how your home and station must be constructed, including electrical wiring. The utility serving your area may also have standards that must be met for them to provide service. Check with the utility and building department to see if they have guidelines or instructions to help you.

AC Power Wiring

As shown in **Figure 3.3**, most residences in the US receive ac power

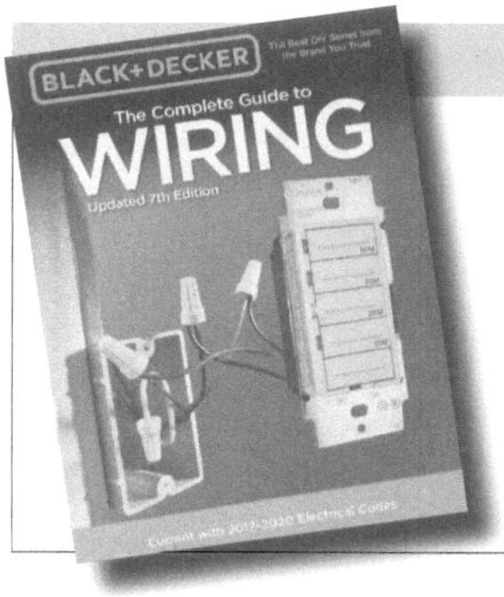

Recommended Home Wiring References

Unless you're a professional electrician, you'll have a lot of questions about how to deal with ac wiring, wherever you live. To help you with all the questions, we recommend a recent title from the series of Black+Decker Complete Guides, *The Complete Guide to Wiring, Updated 7th Edition,* published by Cool Springs Press. The book covers all aspects of home ac wiring requirements, materials, and techniques. It is current with 2017 – 2020 electrical codes and available online for less than $20.

Other recommended books include *Wiring a House* by Rex Cauldwell and *Wiring Simplified* by Hartwell, Richter, and Schwan.

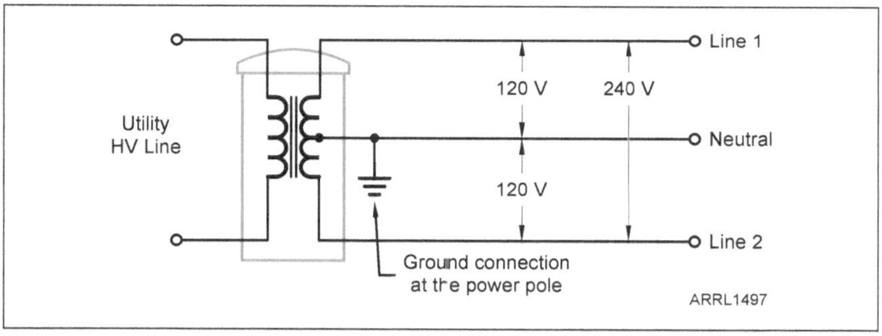

Figure 3.3 — Typical residential single-phase, three-wire ac power service. The electric meter is not shown.

as a *single-phase, three-wire* service. The utility service passes through a meter and then to your residence's circuit breaker box or service panel. (The NEC standard's name for what we call a "circuit breaker box" is a *service disconnecting means enclosure.*) Individual or *branch* circuits connect loads (appliances, lighting, radios) to the service panel. Loads requiring 120 V are connected from *line-to-neutral* and 240 V loads are connected from *line-to-line.*

The 120 V branch circuits have three conductors: hot, neutral, and ground. Loads are connected between the hot and neutral conductors. It is not permitted to connect 120 V loads between the hot conductor and equipment ground.

There are two types of 240 V circuits. The "three-wire" 240 V circuit has two hot conductors (one connected to each line) and a ground conductor — no neutral. Some appliances need both 240 V for heating and 120 V for control and display. They are connected to a "four-wire" 240 V circuit with both hot conductors and ground, along with a neutral conductor, so

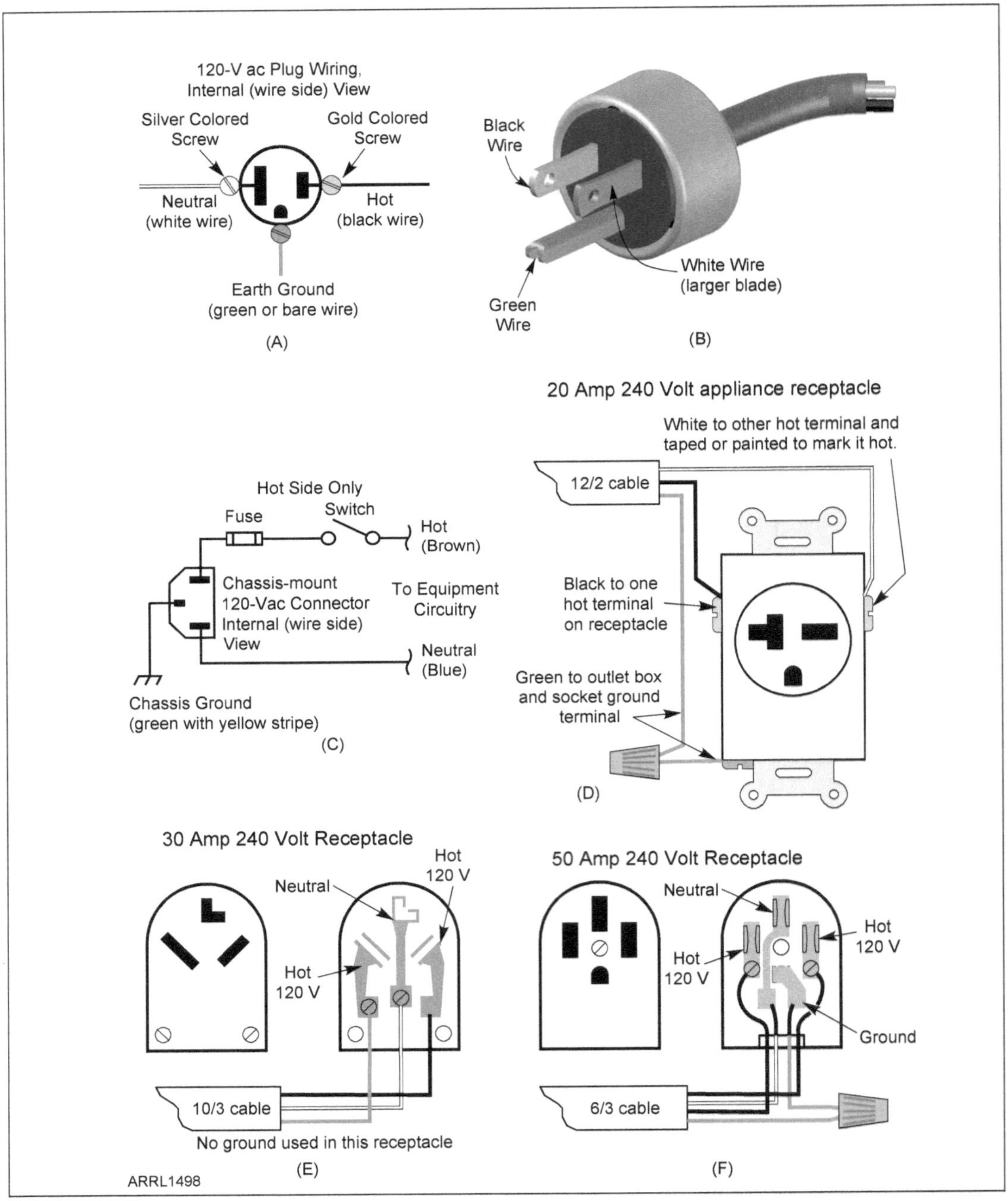

Figure 3.4 — Wiring of common US ac power receptacles: A, B and C – 120 V; D – 240 V, 20 A; E – 240 V, 30 A; F – 240 V, 50 A. See the Appendix for a more complete chart of plugs and outlets.

that 120 V is available. (In most amateur stations, the "conductors" are "wires," but we'll continue to use the standard term.)

You may also encounter older 240 V circuits with three conductors rather than four that use the common as both neutral and ground. **Figure 3.4** shows typical plug/receptacle wiring, and the Appendix includes a more complete drawing showing wiring for 120 and 240 V plugs and outlets.

In the branch circuits, a circuit breaker or fuse (*over-current protection device*, or OCPD) is connected in each hot conductor to open the circuit if there is too much current. This can occur from a short-circuit or a sustained overload.

In a 240 V circuit with two hot conductors, an OPCD must be installed on both hot lines. Both of the breakers in a 240 V circuit are mechanically linked ("ganged") so if one breaker trips, both hot conductors are de-energized.

Circuit breakers *must* be connected in the hot conductor, never the neutral. Although the circuit current flows in both the hot and neutral conductors, opening the neutral leaves the hot conductor energized. This may continue to present a shock or fire hazard through a ground connection. Power *must* be removed from the hot conductor to remove power from the whole circuit.

Bonding for Electrical Safety

Each of the three types of "ground" connections has a purpose in your station. Let's start with the primary purpose of ac safety grounding: to keep you from being shocked. Dissipating lightning strikes and keeping

Why Isn't Ground "Ground"?

The telegraph system was the first wide-area electrical network. The Earth was used as the path by which current returned after flowing through all of the sounders and keys along the circuit. (Search for "Grounding and the Modern Telegraph System" on Mike Holt's website **www.mikeholt.com**.) Later, early dc power distribution and electric trolley systems used connections to the ground in the same way.

Over time, it was learned that using the Earth as a current return path was not a good idea. Soil and rock are not particularly good conductors, so current flowing through them creates a voltage that can be hazardous or cause signaling systems to malfunction. The early days of electrical power saw many electrocutions because we just didn't know how to build systems safely at the time. In addition, the inductance of the return path and its large area resulted in large switching transients and noise.

Without telling the whole history of power safety, today we understand that it is much better to provide a solid, metallic connection to a single, common point in the ac power system. That avoids all the problems associated with electricity in the ground and it also gives current a safe path to follow if there is a fault. Nevertheless, the term "ground" is very old and, once in wide use, it has become impossible to control. Maybe you are starting to see why the word "ground" causes so much trouble!

Why Is Power Cable Called "Romex"?

Like other common brand names, Romex has become a generic reference to a type of cable, but it is a registered trademark owned by Southwire (**www.southwire.com**). The Romex brand of Non-Metallic Building Wire ("NM") originated in 1922 with its development by the former Rome Wire Company, a predecessor to General Cable Corporation, which is now owned by Southwire.

you from getting an "RF burn" are tasks we'll describe in other chapters.

All branch circuits must have a ground conductor that runs back to the service panel. The ground conductor must be at least as large as the current-carrying conductors (hot and neutral or line and neutral) so it can handle the same current overloads without causing a fire hazard. The branch circuit ground conductor is almost always one wire in a multi-conductor non-metallic (NM) cable, usually referred to by its trade name, Romex (see the sidebar, "Why is Power Cable called Romex?"). Inside the cable sheath, the ground wire is usually bare copper, but it can also have insulation that is green or green with yellow stripes.

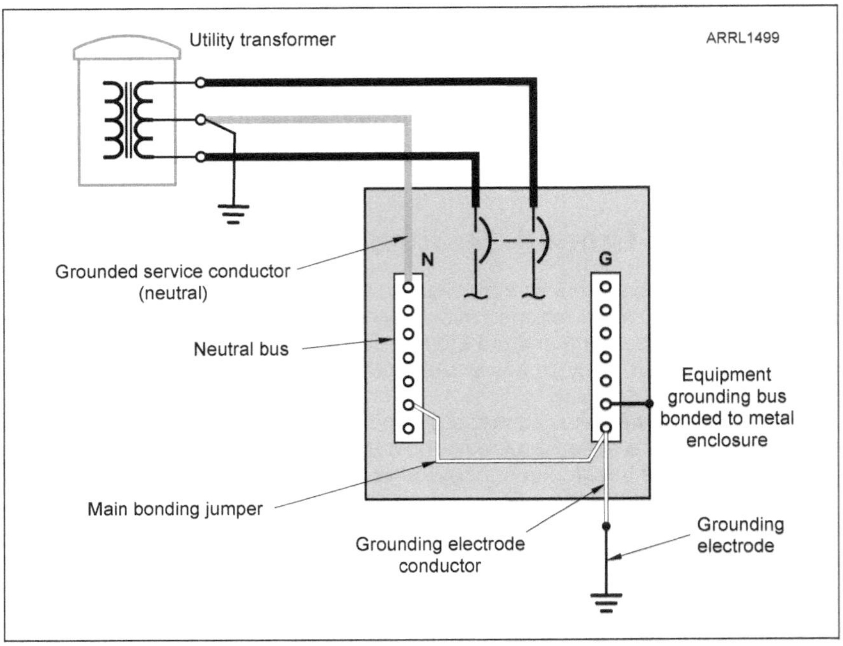

Figure 3.5 — Grounding and bonding arrangement for a single-phase, 3-wire service showing the main bonding jumper between the ground and neutral busses. [Based on information in the National Electrical Code published by the National Fire Protection Association.]

The ground conductors for all of the branch circuits are then connected together on a common ground bus in the service panel. That ground bus is then connected inside the service panel's enclosure to the ac service's neutral bus by the *main bonding jumper*. This is the single, common point for all current paths in the residence. The main bonding jumper must be installed in the service panel, but not in a sub-panel anywhere in the building. Most panels are built with a bonding jumper in place, and it must be removed if the panel will be used as a sub-panel, such as for a ham station.

The branch circuit ground connection is sufficient to protect against shock hazards caused by an accidental connection between either the hot or neutral and exposed metal. The resulting current back to the neutral bus will cause the circuit breaker to trip and disconnect power. This is why it is important to have a safety ground current path in *each* branch circuit with the hot and neutral. (The NEC requires the circuit ground conductor to be run with the hot and neutral conductors to minimize inductance of the current path.) If there are multiple paths, the hazardous current might not flow through the circuit breaker and cause power to be removed. This leaves the hazard fully energized and dangerous.

Grounding the Power System

Building codes in North America require that neutral be bonded to one or more *ground electrodes or earth electrodes* at the service panel. The ground electrodes must also be bonded to the panel, and there must be a bond between the neutral and the panel. For individual residences, this usually consists of a pair of ground rods outside the house, near where the ac service lines are attached. (Apartments and condos also have this connection, but it is made differently depending on the size and nature of the building.) Additional electrodes may be added to the equipment ground anywhere in the system. **Figure 3.6** shows a typical service entrance ground electrode consisting of a pair of ground rods at least 6 feet apart, and many guidelines require the distance to be at least equal to the length of the rod. Electrical codes also limit the length of conductor between the panel and ground electrode.

Ground rods should be 8 feet long, but if you have rocky soil or other special circumstances, check with the local building department for acceptable alternatives. To make a connection to the rod, use clamps and terminals that are *listed*, meaning "approved for the application." Do not solder the connection to the rod. Welding a ground conductor to the rod using "one-shots" works well and is addressed in the next chapter.

Another common type of earth connection is made to the reinforcing steel rods in a concrete footing or slab as shown in Figure 3.6B. This is called a *concrete-encased electrode* or *Ufer ground* (after its inventor).

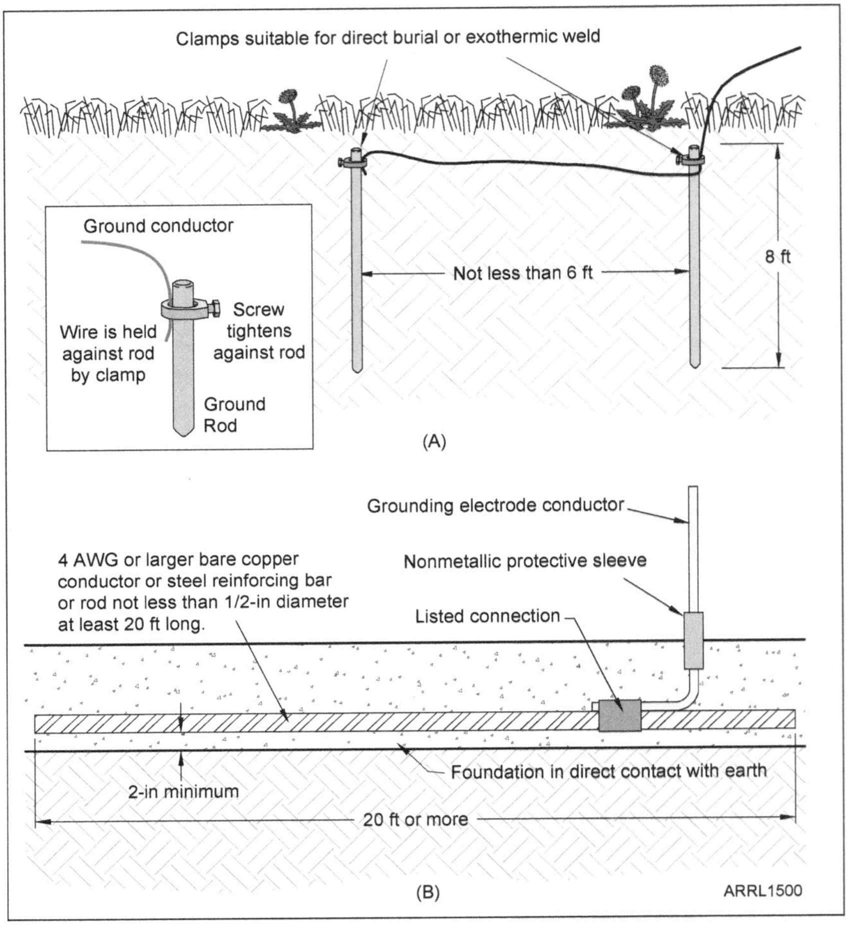

Figure 3.6 — The ground electrode for residential ac service is usually a pair of 8-foot ground rods, 6 feet or more apart, connected together with a heavy bonding wire shown in part A. A concrete-encased electrode or Ufer ground is shown in B. Clamps rated for burial should be used or a welded connection can be made for either type of electrode. The inset shows the proper way to use an acorn-style ground clamp. The conductor must be continuous between the ground rods. [Based on information in the National Electrical Code published by the National Fire Protection Association.]

With the usual large surface area of concrete structures in direct contact with the ground, a Ufer ground makes a very good earth connection. Ufer grounds are discussed more in Chapter 6.

The earth connection is there to provide a stabilizing connection to the Earth in case of a power-line fault and for lightning protection. We'll cover the earth connection in more detail in the chapter on lightning protection.

Call Before You Dig or Drive!

You really don't want to find a buried pipe or electrical service when driving a ground rod! Each state has a toll-free number to arrange an on-site visit to locate and mark utilities. Most are completely free, although it may take a few days to schedule a non-emergency visit, so plan accordingly. Call 811 to be connected with the service for your state or visit **call811.com**.

Check Your Wiring

Before building a station or modifying an existing station, start by making sure your existing ac wiring is in good shape. But before you begin, if you are new to ac wiring or uncomfortable around it, get a professional to do the job, or have an experienced person show you how to do it. Working on ac wiring is safe if you treat it with respect and follow simple safety rules. This is a good time to read the safety section of any wiring handbook or guide, even if you *think* you know what you're doing.

A simple plug-in ac wiring tester will get you started (see **Figure 3.7**) along with a simple ac voltage tester. Each outlet should be checked both electrically and mechanically. If the outlet is cracked or damaged, or if a plug is loose when plugged in, replace the outlet. Even though NEC Section 210-7(d)(3) allows replacing older two-wire, ungrounded outlets without an upgrade, for an amateur station, they should be replaced and the wiring upgraded to provide the ground conductor.

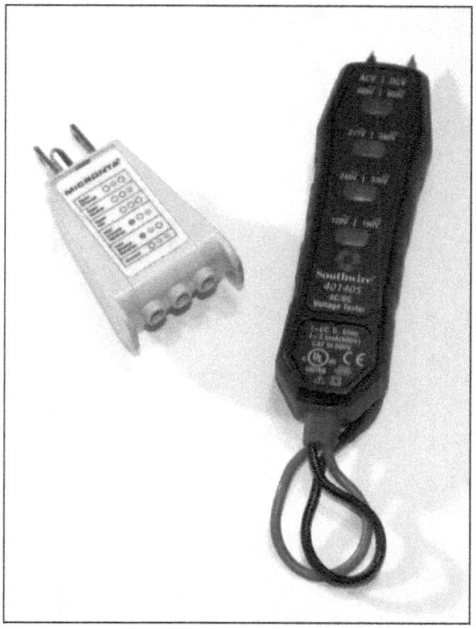

Figure 3.7 — Simple ac voltage tester (left) and plug-in ac circuit tester (right) for checking wiring. Always be sure a circuit is de-energized by using a tester.

Identify which circuits supply power to your station by turning off circuit breakers one at a time with a light or radio plugged into the outlets. Remember that some duplex outlets can be wired with a separate circuit for the top outlet that can be turned on and off with a wall switch. Each room should be on one circuit, but there may be one circuit for lighting, another for the outlets, and so forth. This is a good time to find out if the circuit also supplies another room, particularly if the outlets are on a wall shared by two rooms.

Make a visual inspection of the outlets by removing the cover and looking at the wires going to the outlet. In North America, the neutral is the white wire, and equipment ground wires are green or green with a yellow stripe. Black is used for the hot or line conductor in 120 V circuits; red and black are used for the line conductors in 240 V circuits.

On 120 V circuits, verify that the neutral is connected to the contact feeding the receptacle's longer slot. If anything seems amiss or not quite right, turn off the circuit breaker to the outlet, verify power is disconnected with a tester, then pull the outlet out of its enclosure and inspect it more closely. If there are multiple wires connected together in the box with wire nuts, be sure the nuts are tight and don't show signs of overheating.

The ground conductor must be connected to the designated terminal of the ac outlets. Equipment connects to the ground either as the third pin of a socket or via the center screw of the outlet that holds on the cover plate. If the enclosure for the outlet is metal, the ground wire must be connected to that, as well (see **Figure 3.8**). This connection is most often accomplished by conductive mounting tabs on the outlet that are mechanically bonded to the outlet. The ground contacts are screwed down to the metal enclosure or outlet box.

All ground conductors must be continuous from the service panel to the point at which equipment is powered, such as an ac outlet. Rules about using metal conduit as the ground conductor and how its mechanical junctions are made vary between jurisdictions. Check your local building codes before using conduit in this way.

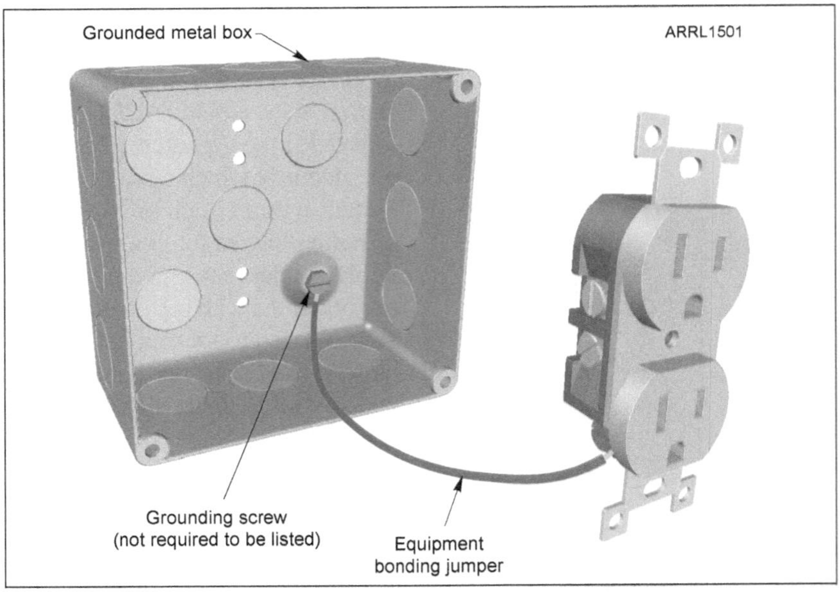

Figure 3.8 — If an ac outlet is installed in a grounded metal box, the outlet ground terminal should be connected to the box. If the outlet ground tabs are not used to make the contact (see text) use a short jumper to a grounding screw or clip. [Based on information in the National Electrical Code published by the National Fire Protection Association.]

Table 3.1
NM Cable Jacket Colors and Current Capacities

Jacket color conventions are voluntarily followed by the manufacturer

White	#14 AWG	15 amps
Yellow	#12 AWG	20 amps
Orange	#10 AWG	30 amps
Black	#6 or #8 AWG	55 or 40 amps
Gray	UF (underground feeder) cable	

This is also a great time to check Ground Fault Circuit Interrupter (GFCI) outlets that feed other regular outlets on the circuit. Make sure the fault detection function works by pressing the TEST button. With the breaker tripped, locate the other outlets on the circuit it protects. In your station notebook (you *do* keep a station notebook, don't you?) write down which outlets are protected by which breaker.

Check the ac service panel internal wiring. The ground wire for each circuit must be connected to a common bus that is tied to the neutral with a main bonding jumper wire or strap (see Figure 3.5). Label the breaker or breakers that feed the circuits going to your station. Verify that the breaker rating is correct for the size of wire in the cable for each circuit, too (see **Table 3.1**).

Power for the Station

A single 20 A, 120 V circuit will provide sufficient power for all of the radio and computer equipment in almost any single-operator amateur station, including most 500 W power amplifiers. A single 20 A, 240 V circuit will provide sufficient power for two legal limit amplifiers that do not transmit simultaneously. A generous number of 120 V outlets should be provided. The 120 V and 240 V outlets should have their equipment grounds bonded together. This simple power arrangement has the major advantage of minimizing problems with audio hum and buzz caused by power system leakage current. Install good quality 20 A outlets and 20 A breakers. Use #10 AWG copper conductors to the service panel to reduce voltage drop under heavy loading. Provide one or more additional circuits for lighting, heating, air conditioners, and other equipment.

3.4 AC Safety Grounding Specifics

Grounding Separate Service Panels

If your station is powered by a separate service panel, such as in a garage or outbuilding, make sure the ground connections comply with your local building code. This is covered by Article 250.32 of the NEC, and two typical situations are illustrated in **Figure 3.9**. See Figure 3.5 and the associated text regarding grounding of sub-panels.

Simple stations can be served by a single branch circuit from the main panel as in Figure 3.9A. For this type of installation, circuit breakers in the additional service panel are not required. You can rely on the circuit breaker in the main panel for over-current protection. The ground bus of both panels must be bonded together. Using the grounding conductor in the cable providing power to the additional panel is sufficient.

Larger stations with multiple circuits must have circuit breakers protecting each one as shown in Figure 3.9B. The additional service panel must have its own ground rod or earth connection, as well. And there must be a bonding conductor connecting the ground bus of each service panel. These rules apply to sub-panels whether they are in the same or different buildings if they are all part of the same electrical service.

The bonding conductor between panel ground busses is particularly important for lightning protection and to insure a reliable, low impedance path for fault current so that breakers will trip quickly. The problem with earth as a path is that it is both a high resistance path and the large loop area of the current path also has high inductance.

Additional Ground Rods

As we will discuss in the next chapter on lightning protection and dissipation, additional ground rods are often installed for panels where feed lines enter a building, for antennas and towers, and a direct connection to the station's bonding bus. All of these external earth connections must be bonded together using heavy wire or strap. **Figure 3.10** shows a typical home with several ground rods — all are bonded together.

For external bonding, a heavy-duty conductor is required. The usual size required is #6 AWG copper, bare (preferred) or insulated. Surge currents from lightning strikes, even indirect, can reach the kilo-amp level. The conductor has to be heavy enough to handle the current without melting, or to conduct for as long as possible before melting. External conductors, especially buried, also have to hold together if they are accidentally dug up or pushed around. The connections to ground rods and utility boxes also have to be heavy enough to hold together and maintain good contact, so use the right materials and parts.

Water Pipes as Ground Electrodes

It is a common practice, long recommended in amateur radio articles and books, to use a metal cold water pipe as a ground electrode. Those recommendations were made in an era when copper and galvanized pipe were used for all water service plumbing. These days, plastic material such as PVC pipe or PEX (crosslinked polyethylene) tubing is the norm, including the buried service between a residence and a water main. There are many tales of discovering that a buried copper pipe leaving a building changes to non-conductive plastic after just a foot or two! Water pipes should never be depended upon for earth connection, but they should be bonded to the system of ground electrodes.

Your local building codes will specify how the pipes inside your

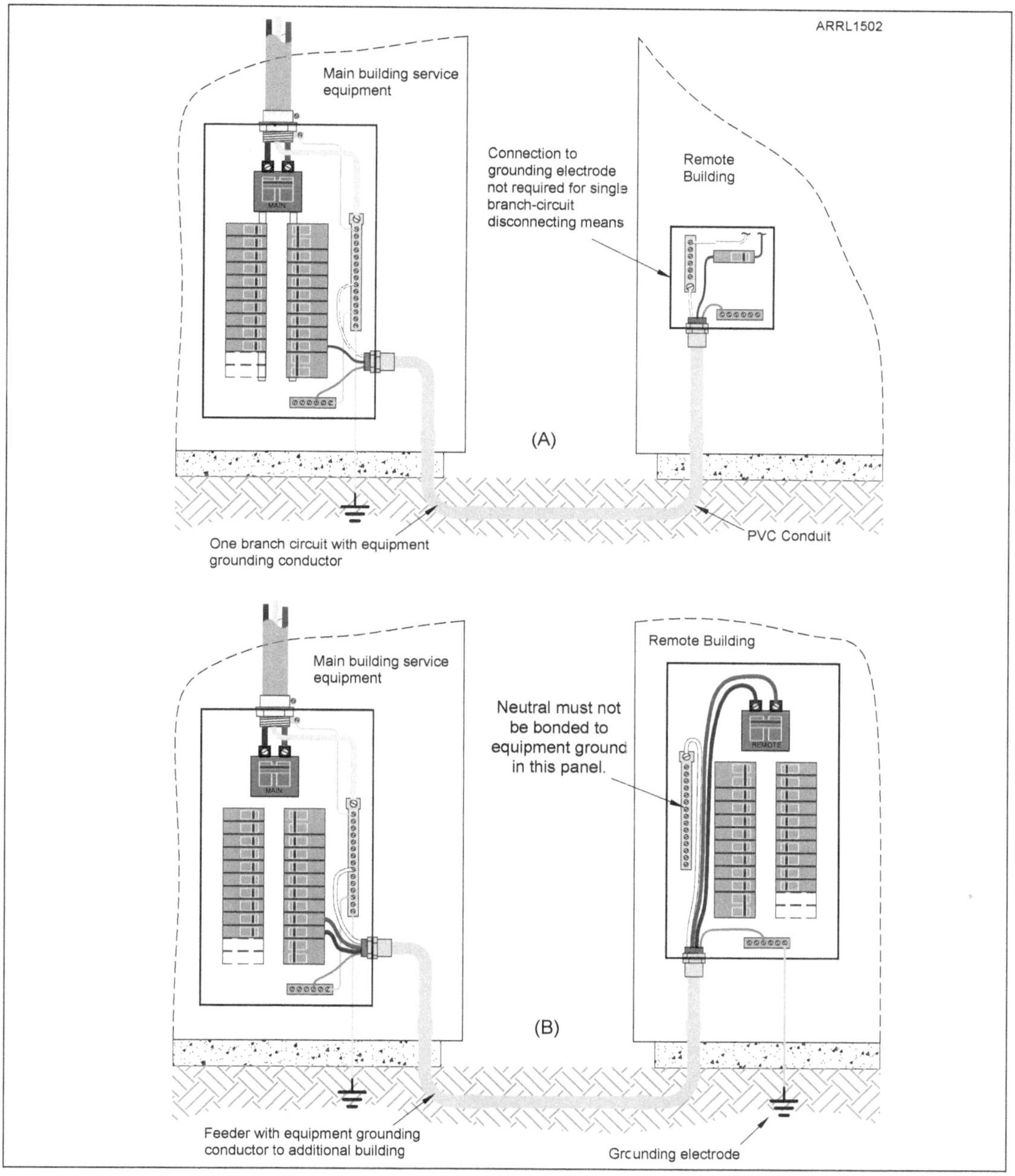

Figure 3.9 — If a separate building is wired with only a single circuit (A), a connection to the ground electrode of the main building's service is sufficient. Most stations will require additional circuits, however. If more than one circuit is used (B), a separate ground connection is required and it must be bonded to the main building ground electrode as well. [Based on information in the National Electrical Code published by the National Fire Protection Association.]

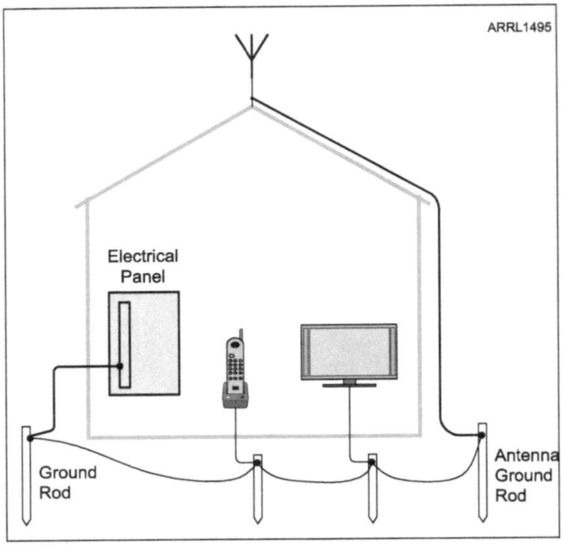

Figure 3.10 — If multiple ground rods are used for different services, they must all be bonded together.

home should or shouldn't be connected together for electrical grounding purposes. (Gas piping should *never* be used for a primary ground connection, although gas pipes may be tied into a home's grounding network.) We'll also touch on the use of pipes for RF management in that chapter. Regardless, always use the right clamps for the type of pipe to avoid corrosion problems that might not become apparent until you have a leak!

Ground-Fault and Arc-Fault Circuit Interrupters

Ground-Fault Circuit Interrupters (GFCIs) are devices that can be used with common 120 V household circuits to reduce the chance of electrocution when the path of current flow leaves the branch circuit (say, through a person's body to another branch or ground). An Arc-Fault Circuit Interrupter (AFCI) is similar in that it monitors current to watch for a fault condition. Instead of current imbalances, the AFCI detects patterns of current that indicate an arc — one of the leading causes of home fires. The AFCI is not supposed to trip because of "normal" arcs that occur when a switch is opened or a plug is removed. **Figure 3.11** is a simplified diagram of a GFCI.

The NEC requires GFCI outlets in all wet or potentially wet locations, such as bathrooms, kitchens, and any outdoor outlet with ground-level access, garages, and unfinished basements — all common locations for amateur stations. AFCI protection is required for all circuits that supply bedrooms. Electrical service into any area with bare concrete floors or concrete masonry walls should be protected by GFCI outlets or breakers.

Older equipment with capacitors in the 0.01 μF to 0.1 μF range connected between ac line inputs and chassis as an EMI filter (or equipment that has been modified with bypass capacitors) will often cause a GFCI to trip, because of the leakage current through the capacitor. The must-trip current is 5 mA, but many GFCIs trip at lower levels. At 60 Hz, a 0.01 μF capacitor has an impedance of about 265 kΩ, so there could be a leakage current of about 0.5 mA from the 120 V line. If you have several pieces of equipment with such capacitors, the leakage current will trip the GFCI. If this equipment must be powered via a GFCI-protected circuit, an isolation transformer can be used to prevent the leakage current from tripping the

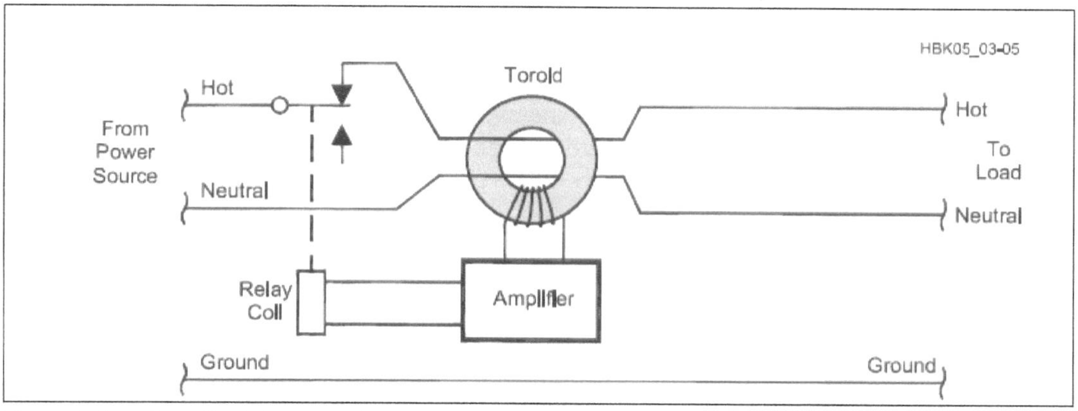

Figure 3.11 — Simplified diagram of a 120 V ac ground fault circuit interrupter (GFCI). Stray or leakage currents to ground unbalance the currents through the toroid which creates a signal detected by the amplifier. A relay then opens the hot conductor, de-energizing the circuit.

GFCI. Be sure all metal parts of the equipment are grounded.

GFCIs and AFCIs are available as portable units, duplex outlets, and as individual circuit breakers. Some early GFCI breakers were susceptible to RFI, but as the technology has improved, fewer and fewer such reports have been received. While it is possible to add filtering or RF suppression to the breaker wiring, a simpler and less expensive solution is to simply replace the GFCI breaker with a new unit less susceptible to RF. For more information on RFI and GFCI/AFCI devices, check the ARRL web page **www.arrl.org/gfci-and-afci-devices**.

3.5 AC Power in Your Station

You've checked all of the wiring and outlets and labeled the circuit breakers and you're ready for business! Except for the fact that you can only plug two cords into a duplex outlet, and most ham stations have about a million things that plug into ac power. Now what? Put away the cube tap and resist the urge to buy a bunch of cheap power strips at the discount store. Let's do this right.

If you have made sure all of your power outlets are properly wired, with a ground connection, then your ac-powered equipment should be properly grounded when you plug it in. There are a few things to watch out for.

Start by making sure your power cords are in good condition and wired according to Figure 3.4. For equipment with a three-wire power cord, make sure that the ground is wired correctly at the equipment and in the plug. If the plug's ground pin has been removed or broken, replace the plug. Use an ohmmeter to verify that the ground pin is actually connected to the equipment's metal enclosure. It is not uncommon for paint to prevent good contact between a ground lug and the enclosure.

Laptop computers and other computer or audio accessories often have two-wire ac power cords and don't have a metal enclosure. Double-insulated appliances and tools also are designed to prevent any single internal failure from creating a shock hazard. These types of equipment do not need an ac safety ground connection to operate safely.

If you use power strips, discard any old strips that don't meet the UL 1363 standard. (This will be displayed on a tag or molded into the body of the power strip.) If the power strip has a surge protector, it should meet UL 1449 (4th Edition), as well. Some strips have lighted indicators to let you know the protective components are still functioning. See the discussion on surge protectors and transient suppression in Chapter 4 for more information on using outlet-connected surge protectors.

You can also build your own power distribution strip with standard outlets in junction boxes. The best way to do this is to use metal outlet boxes and metal conduit between them with all of the metalwork solidly connected together and connected to the ground conductor. You can find an example of this type of project in the Good Practice Guidelines chapter.

Homemade and Vintage Equipment

Take extra care when connecting homemade and vintage equipment to ac power. Both can present significant shock hazards from insulation failures or miswiring. Make a visual inspection, and use an ohmmeter to verify that all connections are secure and there are no short circuits between primary circuits (those connected to the ac line) and the chassis. Replace a two-conductor line cord with a three-wire, grounded cord unless the equipment is an "ac/dc transformerless" type.

When replacing the power cord, the hot conductor should be connected to the circuit containing the power switch and fuse or circuit breaker. (Hot conductors are connected to the smaller blade and neutral to the larger blade of ac power plugs.) The ground conductor should connect directly to the equipment's metal chassis. If the equipment doesn't have a fuse or circuit breaker, add one if practical.

Antique "transformerless" equipment with one side of the ac line connected directly to a metal chassis was designed to operate from either ac or dc power. The way to operate this equipment safely is to power it with an isolation transformer. Ensure that the insulating cover over the chassis is intact and none of the control knobs or switches have set-screws that could contact the metal shaft of a control mounted to the chassis.

Bonding Unpowered Equipment

The typical amateur station includes a lot of accessory equipment that is itself unpowered, but which connects to equipment that is connected to the ac line. Your goal in station building should be to make sure that any exposed metal will not present a shock hazard. This requires you to provide a safety ground connection for any unpowered equipment with a metal enclosure, such as an SWR bridge, antenna tuner, or antenna switch. You can also mount accessories directly to a grounded surface or panel.

Most larger pieces of this type of equipment will have a ground terminal for you to use. If not, add a screw to the enclosure or use a cover or chassis screw for the connection. Be sure to remove any paint from under the screw to insure good contact with the enclosure.

Do Not Use Braid Removed from Coax — Here's Why

It's natural to want to reuse what looks like perfectly good braid from an old piece of coax. After all, it was good enough to conduct RF in a piece of coax, why wouldn't it make a good ground conductor for RF equipment? The braid in coaxial cable is woven around the insulating dielectric of the cable in a continuous process. It is pulled tight against the dielectric as the cable is manufactured, and then covered with a protective plastic jacket. As long as the jacket compresses and protects the braid, all of the braid wires remain clean and in good contact with each other.

Once removed from the protective jacket, however, the braid wires immediately begin to loosen and oxidize or corrode. This reduces the braid's effectiveness at conducting RF quite a bit, making it a poor choice for long-term grounding conductors. To use the coax as a ground conductor, leave the jacket on to protect the braid and treat the coax as a large wire.

The standard for grounding in the communication industry is solid strap or heavy wire. Both can be used indoors or outdoors. Flat-weave, tinned grounding braid can be used if the equipment is subject to vibration or needs to be moved around. Use jacketed braid if it will be exposed to moisture. Do not use braid where it may be exposed to corrosive chemicals.

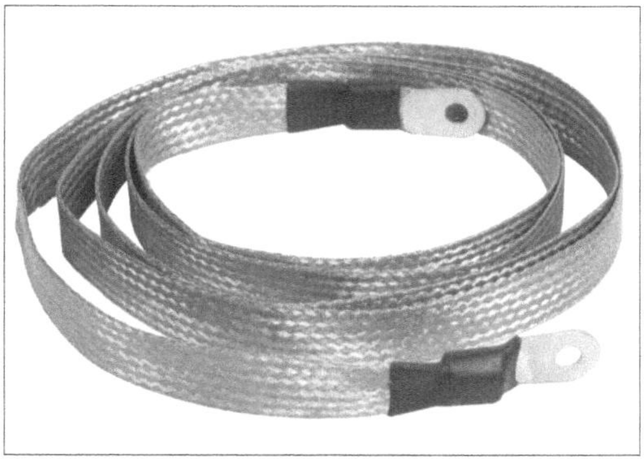

Solid copper strap (A) is the recommended material for grounding and bonding connection at RF. Flat-weave tinned braid (B) can be used if protected from water and corrosive gases or liquids. [Photos courtesy of DX Engineering]

Are Coaxial Cable Shields Okay for AC Safety Grounding?

The question of grounding the shields of coaxial cables to outside antennas is covered in the NEC Section 820.83. But inside the station, you should not depend on a coaxial cable to connect the enclosure of a piece of equipment to the station's ac safety ground system. For example, if you have a multi-position antenna switch, you should connect a separate grounding wire to it along with the coax. After all, the coaxial cable could be disconnected, so you want to have a safety ground for the switch no matter which feed lines are connected.

As you'll see in the chapter on lightning protection, a *bonding bus* is a good way for connecting equipment together and also provides a good way to make ac safety grounding connections. (See the RF Management chapter for a discussion about bonding busses and hum, buzz, and RF current.) Installed behind or under your equipment, the bus can be any heavy metal conductor — copper or aluminum are both available and relatively inexpensive. Strap, pipe, angle, or even heavy wire will do.

Connect the bus to the ac safety ground conductor at a power outlet. Only one connection is needed. The connection can be made to the center screw of an outlet cover (confirm the outlet is properly grounded first) or inside the outlet's junction box directly to the ground terminal on the outlet or to the ground conductor itself. Make a direct connection — don't rely on the power cord of a piece of equipment, which might become unplugged or be removed.

Each enclosure should then be connected to the bonding bus as described in the chapters for lightning protection and RF management. Use wire (#14 AWG or larger) strap, or flat-weave grounding braid. Again, be sure to clean off any paint under the mounting screw on the enclosure.

Rack-Mounted Equipment

Some amateurs mount power supplies, auxiliary equipment, amplifiers, and so forth in a rack cabinet. The equipment may be directly mounted to the rack rails or on shelves in the rack. In either case, section 5.3.3 of the R56 guidelines discusses how to use a *rack ground bus bar* (RGB) similarly to the bonding bus discussed in the preceding section.

The RGB is then connected to the station's ac safety ground. If the rack has its own power cord for an internal power strip, the ground wire of the cord and the ground pin of the ac plug serve as the ac safety ground conductor for the rack. Additional connections for lightning concerns are discussed in that chapter of this book.

Portable and Mobile Generators

It is more common than ever to have a portable generator for camping, emergency, and public service use. The NEC addresses safety grounding for this type of generator in section 250.34. (Fixed generators are not covered here as they are not usually self-installed by homeowners.) The NEC considers "portable" to describe a generator that is easily carried from one location to another by a person. "Mobile" applies to generators that are capable of being moved on wheels or rollers and includes generators mounted in a vehicle.

A ground rod or other direct earth connection is not required for portable generators as long as the generator has receptacles mounted on the generator panel and the receptacles have equipment grounding terminals (i.e. the third ground pin of an ac receptacle) that are bonded to the frame of the generator. Equipment must be connected to the generator through a suitable cord and plug, such as the usual extension cord. Any exposed metal surface of the equipment must be connected through the ground wire of the power cord to the receptacle ground terminal, as well. (See NEC section 250.114.) If the generator is mounted in a vehicle, the same rules apply as long as the equipment supplied by the generator is mounted on the vehicle and the frame of the generator is bonded to the frame of the vehicle. In both cases, it is okay to use a ground rod connected to the generator frame, but you don't have to. (See the chapter on Good Practice Guidelines for more about generators for portable stations.)

UPS and Battery Supplies

Battery power is becoming more common, especially associated with solar power systems and other alternative dc energy sources. For amateur stations, the usual arrangement is for one or more storage batteries in parallel, wired to produce 12 V dc. A maintenance or float charger may or may not be attached to the batteries. An *uninterruptable power supply* or UPS contains one or more batteries and an ac inverter, along with control electronics to switch between external ac power and internal battery power to supply ac output power.

Although the batteries themselves do not produce a shock hazard and are not considered part of the ac safety grounding system, they share a common return connection with much of an amateur station's equipment. Current from a battery supply is as high as from conventional ac power supplies, and fault currents from short circuits can be quite high. The R56 standard recommends at least #6 AWG conductors be used to connect a battery system to a station's ground system. This creates both a connection to an ac safety ground and a high-current path for any short circuits so that the battery supply circuit breakers will disconnect and protect the batteries.

Consult your local building code for rules about installing, maintaining, and protecting a solar or other alternative energy system. The building code may also cover installation of a large uninterruptable power supply (UPS) system. A UPS includes an output transformer, making it what is referred to as a "separately derived system" when it is running on its batteries. Such systems require a bonding conductor between neutral and grounding conductor (green wire) on the output side (done inside the UPS). The green wire must also be grounded, and this may be done via the ac power connection supplying the UPS.

Antenna and Tower Systems

Article 810 of the NEC includes several requirements for safety grounding of antennas and feed lines that you should keep in mind when designing your antenna system. (We will cover these connections in the chapter on lightning protection.) The single most important thing to consider for safety is to address the potential for contact between the antenna system and power lines. As the code says, "One of the leading causes of electric shock and electrocution, according to statistical reports, is the accidental contact of radio, television, and amateur radio transmitting and receiving antennas and equipment with light or power conductors." (See Article 810.13, Fine Print Note.)

The NEC requirements for wire sizes, bonding, and installation practices are mostly aimed at preventing tragedy, by avoiding the contact in the first place, and by mitigating the effects of a contact if it occurs. Article 820 of the NEC applies to Cable TV installations, which almost always use coaxial cable, and which require wiring practices different from Article 810 (for instance, the coax shield can serve as the grounding conductor). Your inspector may look to Article 820 for guidance on a safe installation of coax, since there are many more satellite TV and cable TV installations than amateur stations.

Ultimately, it is the inspector's call as to whether your installation is safe. Article 830 applies to Network Powered Communication Systems, and should you want to install wireless LAN equipment at the top of your tower, you'll have to pay attention to these requirements.

Fiber Optic Service

It is increasingly common for fiber optic service to be run directly to residences (referred to as *Fiber To The Premises* or FTTP). The non-conductive fiber cable will terminate at a service unit that is usually mounted on the side of the building or just inside. The service unit and all wiring inside the building (typically an ac power supply and an Ethernet cable to a router or gateway) must be grounded and wired in the same way as regular network or computer equipment. If the service unit has an external ground rod connection, it must be bonded to other ground rods. FTTP connections are covered by NEC Article 840.

Antenna Bonding (Grounding) Conductors

In general, the NEC requires that the conductors used to bond the antenna system to ground connections for protection against lightning be at least as big as the antenna conductors, but also at least #10 AWG in size. Note that the antenna grounding conductor rules are different from those for the regular electrical safety bonding, or lightning dissipation grounds, or even for CATV or telephone system grounds. Ground connections related to normal antenna operation can be made with #14 AWG or larger wire. This includes connections to radial systems, ground jumpers for feed lines and antenna elements, antenna matching equipment grounding, and so forth.

Motorized Crank-Up Towers

If you are using a motorized crank-up tower, there are also some NEC requirements, particularly if there is a remote control. In general, there has to be a way to positively disconnect power to the motor that is within sight of the motorized device, so that someone working on it can be sure that it won't start moving unexpectedly. From a safety standpoint, as well, you should be able to see or monitor the antenna from the remote control point.

Chapter 4

Lightning Protection

While Mother Nature may have the upper hand as far as when and how much energy lightning delivers, you have the resources and ingenuity to influence where that energy goes. The goal of this chapter is to help you understand the basics of lightning protection.

What Do I Need to Learn From this Chapter?

- The nature of lightning
- How lightning causes damage
- Managing currents and voltages created by lightning
- Using bonding to prevent damage

Along with shock hazards presented by powering equipment from the ac line, the other big safety concern for amateurs is lightning. This is important to amateurs with a fixed station in a residence who must protect their home and station against the effects of lightning. The topic of lightning protection was addressed in a series of articles by Ron Block, KB2UYT (now NR2B), published in the June, July, and August 2002 issues of *QST*. Those articles are the basis for much of this chapter, and they are also available online at **www.arrl.org/lightning-protection**.

These discussions will prepare you for devising a lightning protection plan to create a "zone of protection" for your station. The zone includes all

Operating Safety

No matter how good your lightning protection system is, you must not be in electrical contact with the radio equipment during a lightning strike event. Even assuming there is no current flowing between the radios in your radio room, the voltage of all equipment will be raised above ground potential. If you are touching any piece of equipment, including a microphone, key, or keyboard during the strike event, you may be the path of least impedance from the equipment to whatever ground you are standing on. This ground path can be through the rebar in a concrete floor, to a nearby electrical wire or water pipe, or through unprotected equipment or appliances.

Consider getting a storm warning device capable of alerting you when lightning activity is within 10 miles of your station. (See the sidebar, "Lightning Detectors," later in this chapter.) When the alarm sounds, leave the radio room. If your lightning protection system is designed and installed properly, you may leave the equipment connected and powered on — but you must leave the room and not be near the equipment.

electrically interconnected electronic devices on an operating desk or in an equipment rack. It does not include the antennas, coax feed lines, electrical service both in the station and at the tower, or electronics anywhere else in the residence. (The zone of protection is discussed in more detail in the Lightning Protection Planning section of Chapter 6.)

Many commercial installations are located on hills or mountaintops that are great lightning strike targets. They do survive direct strikes and continue to operate, however. The techniques employed for these commercial sites are described in Motorola R56 — *Standards and Guidelines for Communication Sites*. The level of protection illustrated by R56 is not possible or required for the majority of amateur radio stations, but the principles and techniques in it can be learned and adapted by hams.

4.1 What is Lightning?

We all share the general understanding that lightning is an electrical discharge of considerable size and energy. Even so, it is worth spending a little time getting into the details. While lightning is rightfully intimidating, protecting your station from its effects shouldn't be. By starting from the ground up, you'll understand lightning better and so be more prepared for dealing with it.

Within a thundercloud, the constant collisions among ice particles cause a static charge to build up a large voltage between the cloud and the Earth's surface. Eventually the voltage becomes large enough to cause the electrical breakdown of the air — lightning. The discharge, or *stroke*, can occur within a cloud, between clouds, or between a cloud and the ground, called a *lightning strike*. The strike begins with an initial stroke that establishes the path or *leader channel* for charge to flow. This is followed by one or more *return strokes*, which are the flashes that we see.

When a lightning strike does occur, the return stroke rapidly deposits several large pulses of energy along the leader channel. Air in the channel is heated by the energy to more than 50,000 °F in only a microsecond and hence has no time to expand while it is being heated, creating extremely high pressure. The high-pressure channel rapidly expands into the surrounding air and compresses it. This disturbance of the air propagates outward in all directions. For the first 10 yards or so, it propagates as a shock wave (faster than the speed of sound), and after that as an ordinary sound wave — the thunder we hear.

Many amateurs are fascinated with lightning as a natural phenomenon. It combines many things hams find of interest: weather, electricity, electromagnetic waves, atmospheric and geophysics, just to name a few. More is being discovered about lightning all the time. You can read about ongoing lightning research programs on the National Severe Storms Laboratory website (**www.nssl.noaa.gov/research/lightning**). Lightning has also inspired many works of art, but it is fairly rare that one is actually designed to be hit by lightning! Walter De Maria created *The Lightning Field* in the high desert of western New Mexico in 1977 (**www.diaart.org/visit/visit/walter-de-maria-the-lightning-field**).

Lightning Frequency

Figure 4.1 shows a map of how frequently thunderstorms occur in various parts of the US as calculated by NOAA (**www.weather.gov/jetstream/tstorms_intro**). The probability of having your tower or antenna struck by lightning is governed primarily by where you are located and the height of the tower or antenna. While lightning does strike towers, it is at least as likely to strike or couple its energy to power or telephone lines. This book will discuss protection from both types of events.

The counting criterion is relatively simple — a thunderstorm-day is one in which one or more claps of thunder are heard. This gives us a reasonable view of the country with respect to our exposure to lightning.

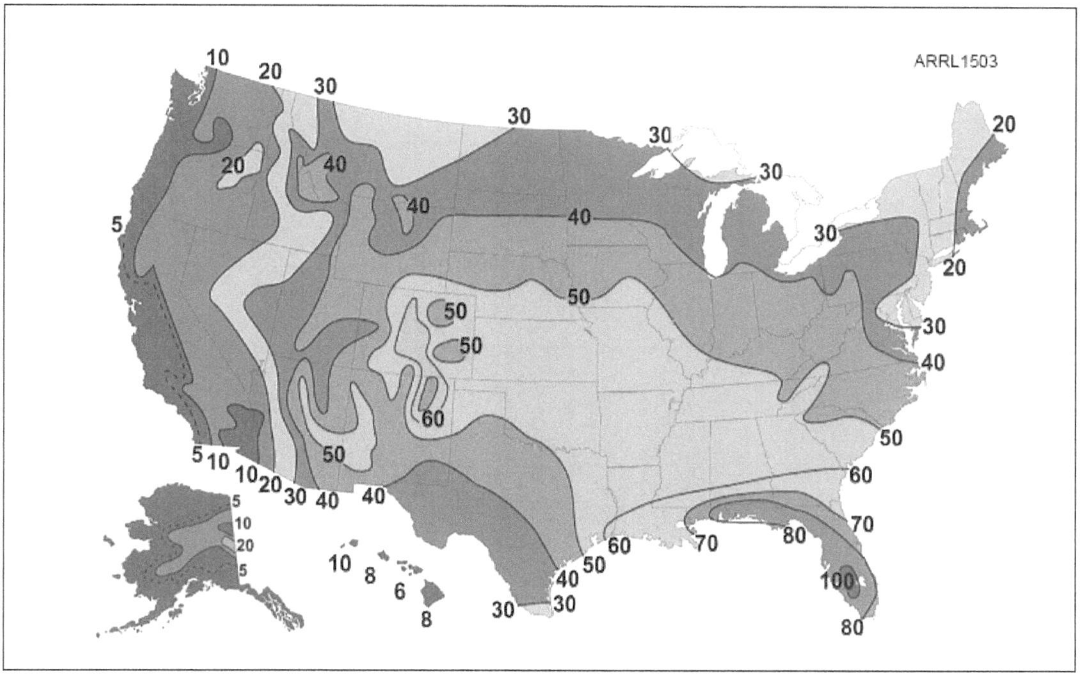

Figure 4.1 — Average number of thunderstorm days each year throughout the United States. The most frequent occurrence is in the southeastern states, with Florida having the highest total (80 to 100+ days per year). [Map courtesy of NOAA]

Each storm generates multiple lightning strikes, so even if you live on the West Coast where storms are infrequent, you should still take lightning seriously!

As you might suspect, the higher your tower or antenna is above average ground level (AGL), the higher the probability of being struck. **Figure 4.2** shows the estimated number of times per year that a tower of a given height would be struck based on the number of thunderstorm-days in your area. Don't forget that even a nearby strike will create substantial currents in wires and cables! You can't assume that because you have a low probability of a direct strike, you can also ignore the hazard entirely.

Lightning as an Electrical Event

We perceive a lightning strike to flicker because, on average, it is composed of three to four impulses per strike. Peak current for the first pulse averages around 18,000 A, or 18 kA (98% of the strikes fall between 3 kA to 140 kA at their peak). For the second and subsequent impulses, the current will be about half the initial peak. The typical interval between impulses is approximately 50 ms. **Figure 4.3** shows a typical impulse, referred to as the IEEE 8/20 model waveform. Remember that this is an average waveform, and that half of lightning strikes have more energy than this waveform!

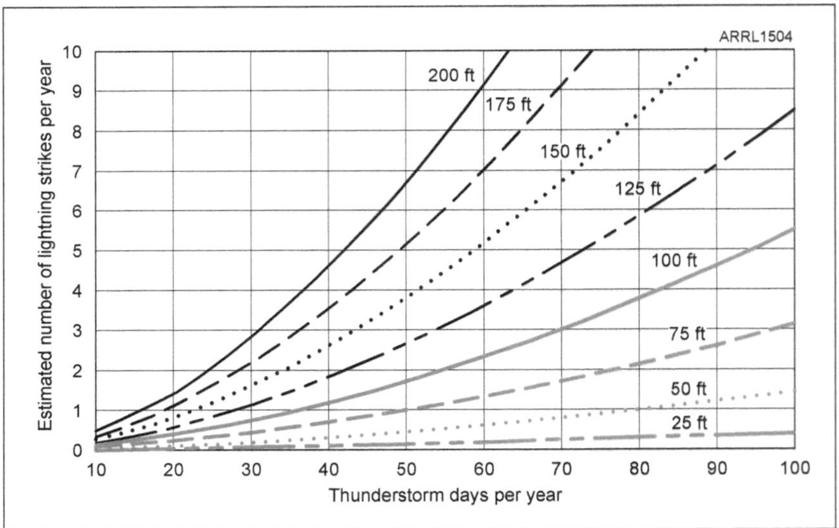

Figure 4.2 — Estimated number of lightning strikes per year based on the number of thunderstorm days in your area from Figure 4.1 and the height of your antenna or tower. [Based on information from *Living with Lightning*, Seminar Notes #ECP-826B Version F, GE Mobile Radio Technical Training, © GE 1985]

Lightning Detectors

While thunder is a pretty strong indicator of lightning in the area, it has a limited range and you have to be listening for it. There are several websites that construct a real-time lightning map from lightning detectors around the world. Two of the best-known online maps are Lightning Maps (**www.lightningmaps.org**) and Blitzortung (**en.blitzortung.org**). These are good websites to monitor if you live in a location prone to thunderstorms.

Along with online resources, you might want to try a standalone lightning detector that can act as an alarm without any relying on an online service. Internet searches will turn up dozens of lightning detectors from pocket-sized units to industrial control system accessories, several of which can activate a relay and even email or telephone if lightning is detected. (You can build one, too; see Radmore, "A Lightning Detector for the Shack," Apr. 2002 *QST*, pp. 59 – 61.) Having a personal detector can be very handy when hiking or camping, too!

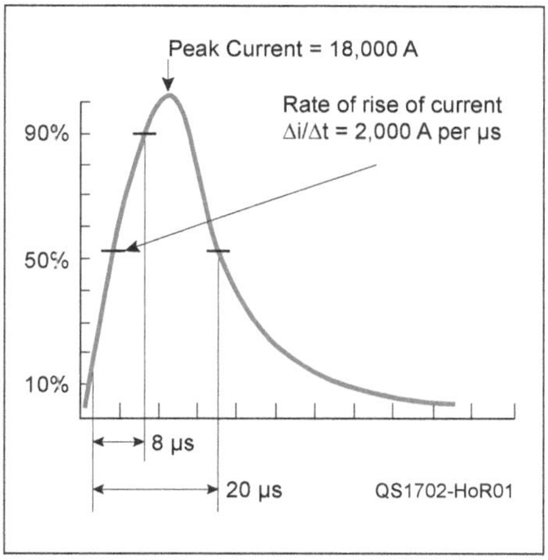

Figure 4.3 — The IEEE 8/20 model waveform for a typical lightning pulse.

The voltage created by this pulse can be enormous and depends on the resistance (R) and inductance (L) through which the current flows. According to Faraday's Law, the faster the current changes through an inductance, the higher the voltage that is created. Higher current through a resistance also means higher voltage, as Ohm's Law states. The combination is:

$$V = (I \times R) + (L \times \Delta i/\Delta t)$$

where the symbol Δ means "the change in," so $\Delta i / \Delta t$ means "the change in current with respect to a change in time" or more simply, "the rate of change of current."

Lightning is a pulse, but most of its energy is between 100 kHz and 10 MHz, with considerable energy all the way to dc. The rise time of lightning strikes is a good indicator of the range of RF energy radiated by the strike. Rise times can vary from a very fast 0.25 μs to a very slow 12 μs, yielding an RF range from 1 MHz down to 20 kHz.

When lightning "attaches" to the ground (when the leader channel reaches the ground) the rise time for current can be as short as 10 ns. This RF content of the strike should be accounted for in the design of your lightning protection plan. In addition to the strike pulses, the antennas and feed lines form tuned circuits that will ring when the pulses hit. This is much like striking a tuning fork in that ringing is created from the lightning's pulsed energy. The result is that lightning's energy extends to 10 MHz with significant harmonic content as high as 100 MHz.

You will sometimes see lightning given as the cause for *electromagnetic pulse* (*EMP*) events. While there are some similarities between the two, lightning is relatively slow compared to EMP. Lightning is primarily a current transient or pulse. EMP events create large voltage transients or gradients that act 10 times (or more) faster than lightning. The damage from both can be similar, but protection techniques are different. EMP protection is not covered by this book but you can read about it in the *QST* articles by Dennis Bodson, W4PWF, which are online in the ARRL's *QST* archives.

Damage from Lightning

What causes the damage when lightning strikes? Is it the voltage or the current? The answer is "Both!" Let's say the average lightning current pulse of the previous section is flowing through 1 foot of #12 AWG wire, a pretty good conductor. The resistance is 0.0016 Ω and the inductance is 0.35 μH. The resulting peak voltage, $V = 18{,}000 \times 0.0016 + 0.35 \times 2000 = 728.8$ V. That's *per foot* of wire! Just a 6-foot length will have 4.37 kV from end to end, even if it is completely straight.

That voltage is why inductance is more important than resistance in bonding for lightning protection. Obviously, your goal as a station builder is to make sure as little as possible of that current flows through anything in your station! (We'll address inductive coupling between the strike current and your station later.)

The large currents in a lightning strike cause mechanical damage from heating, often simply vaporizing wires and PC board traces. In addition, large currents also create large magnetic fields. Those fields are quite powerful and are capable of bending or breaking the conductors they flow through. How powerful? If two conductors are 1 cm apart, each carrying the typical 18 kA current pulse, the force pushing them apart is about 500 pounds *per foot*. You can imagine how brackets and clamps might be broken or ripped free. Current at this level will expand any loops in conductors and can easily break cables or wires. These two effects, heating and bending, are important reasons why lightning protection conductors are required to be so large, and why crimp or compression connections are required. A protective connection isn't going to do much good if it melts or breaks. Use materials that are adequately rated for lightning protection and have been shown to do the job!

Those same large currents can also induce a lot of energy into conductive loops near the tower or other objects carrying the main strike current. If a large, open loop of conductive material lies within the field produced by a direct strike, voltage can build up across the ends of the loop until arc-over occurs. It is not unusual to have arc paths of 2 feet in length. The arc may cause trauma to humans and damage equipment.

Another way lightning causes damage is through a shock wave created by the ionized channel and from vaporized material. We hear the shock wave as thunder as the shock wave spreads through the surrounding air. If the channel goes through water-bearing material, such as a tree, some of the water is vaporized into steam, blowing off bark or destroying the tree outright. If the stroke goes through an antenna, it is not uncommon for parts of the antenna or feed line to "disappear!" This is another reason to follow the protection guidelines and use a suitably heavy conductor that won't fail when it's most needed.

Hopefully, the main channel for the discharge won't go through your house or station! Induced currents from a nearby strike or part of the main stroke can still make trouble, though. An arc can develop due to the momentary voltage differences between pieces of equipment — this is another reason why bonding is so important. The arc acts as a mini-strike, causing currents that damage cables and electronics, accompanied by a miniature thunderclap. The voltage induced in cables due to a large current can damage or destroy sensitive electronics, even if there is no visible damage.

The Earth as Ground

It is easy to think of the Earth as one big sink into which current disappears, but the reality is quite different. There are many types of soil, with different electrical characteristics. Resistivity (measured in ohm-cm with lower values being more conductive) or conductivity (measured in millisiemens (mS)/m) is the characteristic of most interest in this case.

When the current from a lightning strike enters the soil through a ground rod or other ground electrode, it spreads out in all directions, depending on how well the soil in its path conducts electricity. The current density is highest where the electrode first contacts the soil (such as the top of a ground rod) and drops off with depth in the soil and with distance from the electrode.

Common soil types range from dry sand and gravel to a rich "gumbo" with resistivities that vary by a factor of several hundred. Even within a single type of soil, resistivity can vary under different conditions by more than a factor of 100, for example between wet and dry periods or warm and cold periods. As a result, the currents and voltages from lightning strikes will also vary by similar amounts.

Because soil conditions vary so widely, the lengths, sizes, and spacing for ground rods and other types of ground electrodes have been standardized to work in a wide range of conditions. The best approach for hams with limited resources is to follow these recommended practices.

Even if the lightning's current pulse is completely contained within the ground, it will create large voltage differences between points on the ground. This is due to the resistance of the soil. If two ground rods are separated by several feet or more and not bonded together, lightning-sized currents flowing between them can create substantial voltage differences. Different pieces of equipment connected to these "grounds" will exhibit that same voltage difference! This is why it is important to bond together all earth connections — to minimize the voltage differences between them and the equipment connected to them.

There are two basic ways to mitigate damage from lightning strikes. First, create low-resistance, low-inductance paths to steer as much of the current as you can to the Earth where it can be dissipated safely. Second, for the current that does flow in your home and station, minimize the voltage differences that damage equipment and create arcs. Let's start with current paths.

4.2 Managing Current Paths

You can't really "control" lightning, but the easier you make it for the strike energy to dissipate in the Earth instead of heading toward the station, the less your equipment protection will be stressed. Just like "regular" current, lightning divides between all available conducting paths. That's the primary purpose of the external ground system — to provide a low-impedance path to the Earth in order to disperse as much of the lightning energy as possible outside and away from your station. Recognizing the limited resources available to most amateurs, let's see what can be done.

Tower and Mast Ground Systems

Figure 4.4 shows an example of a commercial-style tower ground system. The ground ring shown in Figure 2.2 in Chapter 2 is another good method. Both create the necessary current paths. Treat any permanent conductive antenna support as a tower for the purposes of this section.

When possible, a tower should be separated from the building by at least 20 to 50 feet. By placing the tower at some distance from the building, you minimize the amount of magnetic energy that it will couple into the wiring. In addition, you take advantage of the inductance of the feed line in limiting the surge energy headed toward your equipment. This allows more time for the tower grounding system to absorb the strike energy. If your tower or mast is adjacent to the house, take extra care to create an effective ground system.

Non-conductive antenna support masts such as wooden poles should have a separate ground conductor running from the top of the pole to a

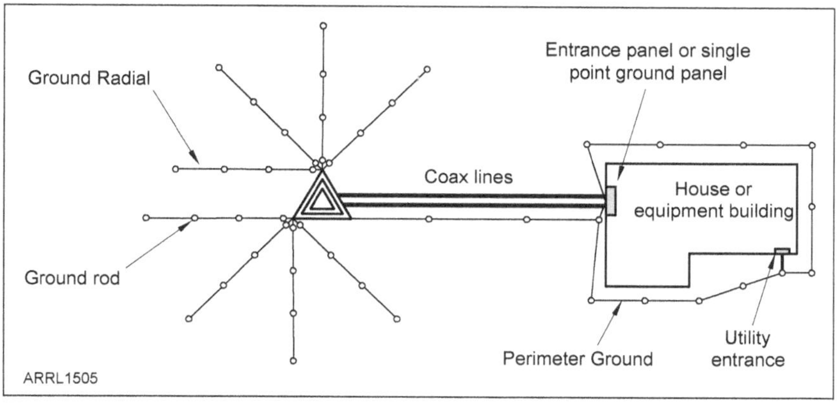

Figure 4.4 — Plan view of tower and equipment building showing recommended ground radials and perimeter ground. The ground ring in Figure 2.2 is an alternate method of grounding the tower. See Figure 4.6 and the associated text for a discussion of perimeter grounds for buildings. [Drawing from *The "Grounds" for Lightning and EMP Protection*, Roger R. Block, PolyPhaser Corporation, Minden, Nevada]

ground system at the base. The wire or strap can be attached to the pole with staples or clamps.

Radials and Ground Rods

Spreading out from the base of the tower is a set of radials. (Eight are shown in Figure 4.4 but for amateur installations, three radials as in Figure 2.2 may be sufficient as a basic system. If you can, install more radials.) Each radial is a bare copper wire or strap buried 6 to 18 inches below grade. The radials should be positioned so that the energy is dissipated away from the residence. While the number of radials required for a particular installation will be dependent on the soil conditions in your location, the system shown here is a reasonable start.

Connected to the radials are ground rods. The ground rods are spaced approximately twice the length of a ground rod. For an 8-foot rod, the spacing would be 16 feet. During the strike, each ground rod has a cylindrically shaped region of influence centered on the ground rod. This is the region in which the ground rod disperses the strike energy. If the ground rods are placed closer together, hoping to lower the impedance of the ground, the regions of influence begin to overlap and the ground rod's ability to disperse energy will be slightly diminished. It is best to stay with the recommended spacing and length of ground rods for the most effective ground system.

While a very long radial/ground rod system is good, electrical and economic considerations must be considered. Analyzing the average cost

Connecting to a Ground Rod

Ideally, the connection to a ground rod should be made using an exothermic bonding process, usually thermite welding. (This is considered a welded connection.) This connection will most likely outlast the life expectancy of the ground system and you won't have to do annual inspections. (Ground rods and Ufer grounds are also discussed in the chapters on AC Safety Grounding and Bonding and Good Practice Guidelines.)

A number of manufacturers supply the molds and fusing material for a variety of cable/strap and ground rod sizes: Erico Incorporated (**www.erico.com**), Harger (**www.harger.com**), and Alltec Corporation (**alltecglobal.com**). You can obtain the necessary items from local electrical suppliers — be sure you get the correct sizes for your ground rod and ground wire. KF7P Metalwerks has produced an informative video on making this type of connection at **www.youtube.com/watch?v=T5DoB26TFtI** or search for "KF7P Uni-shot video" online.

If the exothermic process is not used, mechanical clamps that can be used to connect the radial to the ground rod are available. Be sure to use clamps and hardware that are designed and rated for ground connections, particularly if the connection will be buried. All mechanical connections must be inspected annually or more frequently to ensure the integrity of the system.

of installing an additional foot of the grounding system versus the benefit of a lower impedance system, the break-even point is somewhere around 80 feet. For practical purposes in areas with reasonably conductive soil, the maximum length of a radial should be limited to approximately 80 feet. If the impedance of the ground system needs to be lower, additional radials should be used as opposed to longer radials.

Some caution must be exercised when laying out the radials. If a radial comes within 4 feet of a metal object, it must be bonded to the metal object. The 4-foot rule applies to objects that are above, below, to the left or to the right of the radial. This includes metal fence posts, children's swing sets, buried fuel tanks, and so on. This is to prevent arcs or hazardous voltages from developing between the radial and other nearby metal objects. (Check your local building code or the NEC for information on bonding to fuel or gas equipment and storage tanks.)

Be sure to use the proper techniques for connecting dissimilar metals, such as when connecting the ground system to the tower. Copper should not be connected directly to galvanizing or aluminum. Most towers are zinc-coated steel (galvanized). Connecting a copper wire or strap directly to the tower leg will cause the zinc to erode, allowing the base steel to oxidize (rust). This in turn will increase the resistance of the connection and, over time, may threaten the mechanical strength of the tower segment.

One solution to the corrosion problem is to use a buffer layer of stainless-steel shim stock between the zinc and the copper. (See the following section on Spark Gaps for a photograph illustrating this technique.) Stainless steel shim also works when connecting copper to aluminum, such as tower legs. Several manufacturers offer tower leg clamps for connecting copper to galvanized surfaces. Search online for "tower leg clamps" or "tower base clamps."

If you are constructing a new tower, you can use the tower base itself as a Ufer ground electrode (see Chapter 6). This technique utilizes the rebar that reinforces the concrete base as a ground connection. The rebar itself must be electrically interconnected so there are no spark gaps between the rebar segments, and there must be at least 4 inches of concrete between the rebar and the surrounding earth. If this is done, a wire can be brought out from the rebar and attached to the tower leg. Another option used for slab-foundation dwellings is to put at least 20 feet of heavy copper wire in the base concrete, bonded to the rebar, with wire extending from the base as a ground connection. Rebar and other connections for a Ufer ground must be done correctly — if you are not familiar with these practices, hire a professional to do it for you.

Contrary to some stories, a lightning strike to a properly constructed Ufer ground will not blow up the concrete. Since you must put rebar in the concrete anyway, use it to augment your ground system. Damage to concrete from lightning is usually caused by improper rebar bonding, which allows a spark gap between segments to instantaneously heat the water vapor in the concrete. The resulting expansion causes the damage.

Here are some general guidelines for adapting this to your specific situation.

- In general, doubling the number of radials lowers the impedance of the ground system by one half.
- Radials don't have to be straight; they can follow the contour of your property or flow around obstacles. Make turns gradually (12-inch radius or larger) to minimize inductance created by the bend.
- Short ground rods (even as short as one foot) are better than none at all. Just place them closer together, spaced at approximately twice their length
- If you are fortunate enough to have multiple towers, each should have its own ground system of radials and ground rods. If the towers are within 100 feet of each other, ground rods and one radial should be used to interconnect the towers.

Radials for RF Ground Screens

Is it possible to use radials for both lightning protection and as a ground screen allowing the tower to be used as a vertical antenna? Not really. The radials that create a lightning protection ground system should be buried by 6 to 18 inches. This is too deep for them to be effective at RF where the radial wires are typically laid on the ground or buried no more than a few inches at most. The radials for both systems should be connected together at the base of the antenna, and the RF ground screen will provide some additional lightning protection.

AC Safety versus Lightning Protection Ground Systems

What is the relationship between these two systems? Do they complement or contradict each other? Both have similar purposes — to provide a safe path for current that would otherwise be hazardous. Done properly, the systems need not compromise each other.

The safe path for current in the ac safety ground is back to the common connection between the power system neutral and the ground conductor. The safe path for current in the lightning protection system is away from your equipment to the ground electrode and then into the Earth. As long as all of the ground electrodes of these two systems are bonded together, they do not interfere with each other and can even provide additional safe current paths.

The two systems must be bonded together at their common points — the ground electrodes. You cannot rely on the lightning protection system alone to provide ac safety and vice versa. If properly bonded, ac fault current can flow through the lightning protection ground connections back to the service entrance ground electrode and then to the neutral bus in the service entry panel. Similarly, the shared ac safety ground connections help keep all of the equipment at the same voltage so that lightning-generated currents flow into the Earth instead of between pieces of equipment. The bonding jumpers are key to both systems working to protect you and your equipment.

NEC Article 810 Grounding Rules

Article 810 of the NEC covers TV and satellite antennas, shortwave and AM broadcast receiving antennas, as well as amateur radio and Citizens Band antennas. A tutorial by Mike Holt on the complete Article 810 is available at **www.mikeholt.com**.

The most important sections of this article for lightning protection concerns are summarized as:

810.15 — Outdoor masts and metal structures that support antennas must be grounded.

810.20 — Each feed line or lead-in conductor from an outdoor antenna must be provided with a listed *antenna discharge unit*. (Attic or other indoor antennas are not required to have an antenna discharge unit.) The antenna discharge unit must be located nearest the point of entrance to the building, but not near combustible material. The antenna discharge unit must be grounded. If an SPGP is used, the antenna discharge unit should be mounted on it, whether it is inside or outside the structure. (Hams usually refer to an antenna discharge unit as a *lightning arrester* or *lightning protector*.)

810.21 — Grounding conductors used for lightning protection must be at least as big as any antenna lead-in but no smaller than #10 AWG copper, #8 AWG aluminum, or #17 AWG copper-clad steel. (Ground conductors that are only part of the transmitting system may be as small as #14 AWG.) The conductor may be insulated or bare (preferred) and it must be secured in place. The conductor must be run in a straight line as much as possible and terminate at the nearest available grounding electrode using listed clamps or a welded connection. If the grounding electrode is

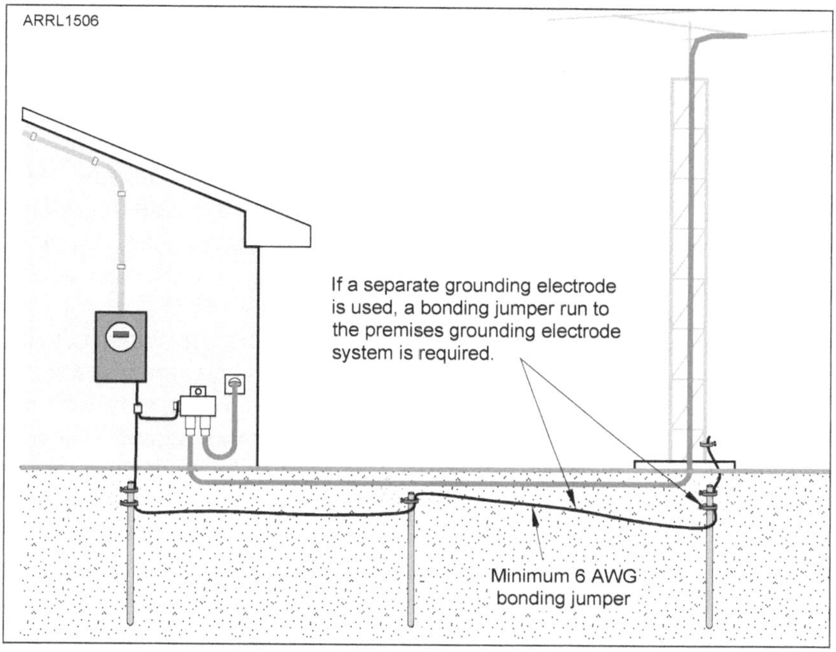

Figure 4.5 — The bonding jumper between an antenna system ground electrode and a building's ground system must be at least #6 AWG. [Courtesy Mike Holt Enterprises of Leesburg, Inc.]

separate from the building's ground system, it must be bonded to the building ground system with at least a #6 AWG conductor as shown in **Figure 4.5**. Don't rely on feed line shields to act as protective ground. (Figure 4.5 is only intended to illustrate jumper size requirements and does not include an SPGP.)

4.3 Bonding to Equalize Voltages

Even though your antenna system may be properly grounded and voltage transients clamped or dissipated, a lightning strike is just too big. Whether the strike is on your tower, to the power lines that supply your home, or even just nearby, your equipment will experience a significant voltage surge as the energy makes its way to the Earth. During a strike, lightning energy will flow toward anything that is capable of being an energy sink. This includes any parts of the system that are not at the same voltage.

Not only will there be a voltage surge, but the different paths the energy takes through your antenna system, station, and home will create large voltage differences. Because electrical signals travel at about 1 nanosecond per foot, fast rise times of lightning transients can create significant voltage differences for short times due to travel differences.

Lightning-induced voltage differences cause damaging secondary arcs and currents — unless you take steps to equalize the voltage throughout your system. That's the purpose of the perimeter ground and of bonding.

Outside your home, follow the requirement to create a perimeter ground and bond all ground electrodes together for both ac safety and lightning protection. This ensures that all of the various ground points in the system are as close to the same voltage as possible during a transient or surge. It also allows all of the ground electrodes to share the current pulse and spread it out.

Inside your station and home, bonding is the last line of defense against voltage transients. Bonding minimizes the voltage difference between equipment enclosures, reducing the likelihood that the voltage is great enough to cause destructive equipment failure. Even a short voltage pulse lasting a few microseconds can damage the sensitive electronics that are common in today's communication and consumer gear.

Perimeter Grounds

Bonding ground rods and other ground electrodes such as water pipes and rebar in concrete footings and bases is the first step in creating a protective ground called a *perimeter ground*. (This is also known as a *peripheral ground* and sometimes a *ground ring*, not to be confused with the ground ring shown in Figure 2.2.) The perimeter ground consists of several ground rods (or other ground electrodes) bonded together with a heavy wire or strap buried at least 6 to 18 inches. Burying the bonding conductor increases the perimeter ground's contact with the soil and protects the conductor from damage.

The perimeter ground should go completely around the building, following the approximate outline of the building, and be anywhere from a few to twenty-plus feet from the foundation. The actual distance from the foundation will depend on each individual situation and what is in the immediate area. As a general rule, porches, fuel tanks and swimming pools should be on the inside of the perimeter ground, while towers and utility transformers should be on the outside.

While the perimeter ground should extend all the way around the building, a perimeter ground that only goes three-quarters or halfway around the building is better than no perimeter ground at all. Although flowerbeds, walkways and driveways frequently present insurmountable obstacles, do your best to get most of the way around. Do not run the perimeter ground conductor under a building or through a crawl space.

If you find that you need to run the grounding conductor along or over some of the building structure, section 250.64(B) of the NEC requires you to securely fasten it (or the conduit protecting it) to the structure, whether that is concrete, wood, or other material. (Aluminum wire is not permitted to contact masonry directly.) If the wire is smaller than #6 AWG, it must be protected from damage by conduit.

The perimeter ground serves two purposes. First, it helps conduct the surge energy around the building, minimizing the ground potential differences under the building during the strike event to minimize lighting currents that might flow through the building. Second, it enhances the basic ground system by providing more connection points to the Earth. The existing ac service entry ground is also connected to the perimeter ground — this is a must!

With a perimeter ground (even partial) in place, the next step is to be sure it is bonded to any other nearby metal objects, including generator and solar power systems, fences, air conditioners or heat pumps, fuel tanks, and so forth. (This is covered by NEC Article 250 and other "special wiring" requirements an electrician will be familiar with.) Check with an electrician or electrical inspector for any special rules that apply regarding grounding fuel lines and tanks.

Figure 4.6 illustrates how a perimeter ground is approached for a commercial installation. Note that the tower has its own ground ring and radials. The building's perimeter ground is connected to the tower's ground system along with the generator system and a nearby fence. Your

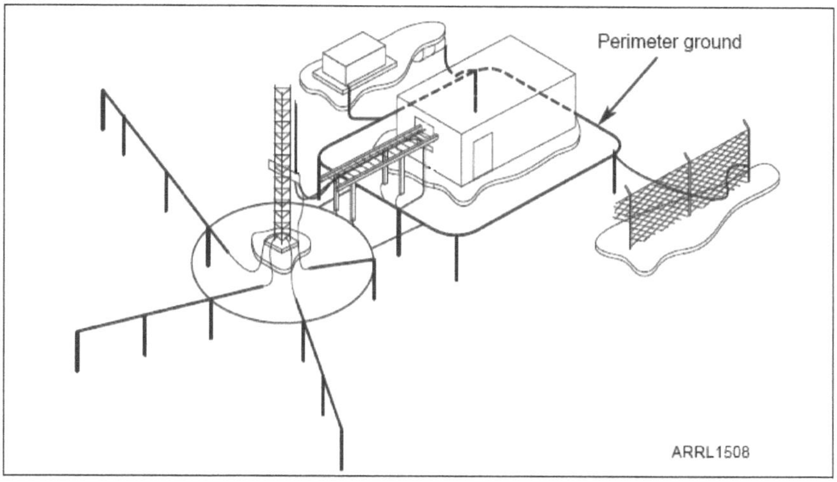

Figure 4.6 — A perimeter ground system for a commercial installation showing how it is bonded to all ground electrodes and nearby metal structures and equipment. Amateurs should run cables on the tower all the way to the ground, if possible. [Drawing from *Standards and Guidelines for Communication Sites*, R56, courtesy of Motorola, Inc.]

station will probably not be constructed to these standards, but it is useful to see how commercial installations are built. Club repeaters or wireless data nodes are often installed in this type of facility, so it is good to understand how they are constructed. (Section 1.2.3 of MIL-HDBK-419A includes design guidelines for several types of perimeter grounds, which it refers to as "ring grounds.")

The bonding conductor that forms the perimeter ground should be non-insulated and buried to act as an additional ground electrode. Be sure to use ground clamp hardware that is rated for burial, or use welded connections. All of the services and cables that cross the perimeter ground should be protected to divert current and voltage surges to the perimeter ground. If you have a roof-mounted antenna, its ground conductor should be bonded to the perimeter ground, as well.

Figure 4.7 shows the usefulness of bonding ground systems together with a perimeter ground. A typical home and station without a perimeter ground are shown in Figure 4.7A. A lightning strike on the tower or the power line may be partially diverted to the Earth through the ground electrodes. Unfortunately, there are other paths for it to follow that go through the home and station. In fact, without a perimeter ground bonding the ground electrodes together, the better you ground your amateur station and antenna system, the more likely it is that lightning transients on the power line will travel through the station on their way to the Earth!

In Figure 4.7B, a perimeter ground has been added along with an SPGP (see section below), bonding the station lightning protection ground to the ac service entry ground. Additional ground rods have been added between the two ground connections. Now, the energy from the lightning has a much lower impedance, direct path to the Earth that does not go through the station. The ac power protector mounted on the SPGP also routes lightning energy away from the circuit supplying the station to the perimeter ground.

These drawings do not address the voltages and currents that lightning induces on internal wiring, both power wiring and wiring for various low-voltage systems (telephone, wired Ethernet, CATV, home entertainment systems). These transients can reach 3 kV, requiring protection on internal branch circuits, as well.

Coaxial Cables on Towers

Depending on the height of the tower (inductance increases with height) and the maximum current in the lightning strike, the tower could easily have an instantaneous voltage difference between the top and the ground that exceeds 100 kV during the strike! If everything on the tower is not bonded together, cables and antennas can easily be damaged. It is a

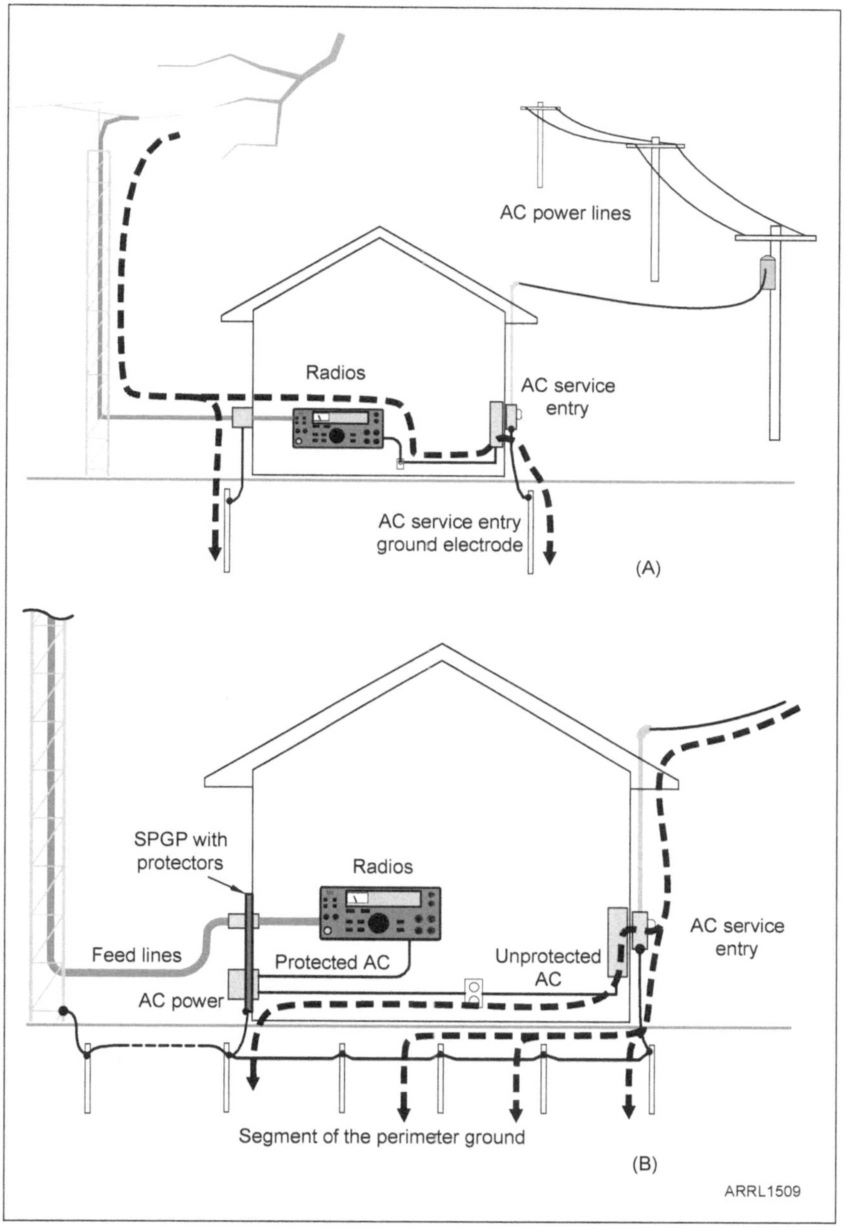

Figure 4.7 — At A, without a perimeter ground, lightning striking the power line or tower will take paths through the building to reach additional ground electrodes. A perimeter ground and SPGP has been added at B, protecting feed lines and ac power, and providing low-impedance paths to the Earth, minimizing lightning energy in the station and home.

good idea to add cable inspection for possible lightning damage to your regular tower and feed line maintenance. Each coaxial cable traversing your tower should be bonded to the tower, as shown in **Figure 4.8**. The first point is at the top of the tower. If one terminal of the antenna connection, such as a coaxial connector or an all-metal J-pole, is securely connected to the tower (referred to as a "dc ground" connection) that can serve as a bonding point. The second point is where the coax leaves the tower to go to the radio equipment. This take-off point should be as close to the base of the tower as physically possible as illustrated by Figure 4.8.

In simplified terms, if the tower is viewed as if it were a very long inductor with one end connected to ground and the coax take-off point as a tap point on that inductor, you can begin to appreciate the problem caused by allowing the coax to leave the tower at any point above the bottom. For a 100-foot tower with a coax take-off point at the 10-foot level, the coax shield will rise to about 10% of the top-to-bottom voltage along the tower.

If the antenna is not connected to the tower, such as for an insulated driven element, bonding must be done where the coax begins to run down the tower by using a coax grounding kit such as in **Figure 4.9**. A ground kit should be installed at the top of the tower to minimize strike current through the coax connector shell. If a ground kit is not used, the high strike current can weld the shell of the coax connector to the antenna con-

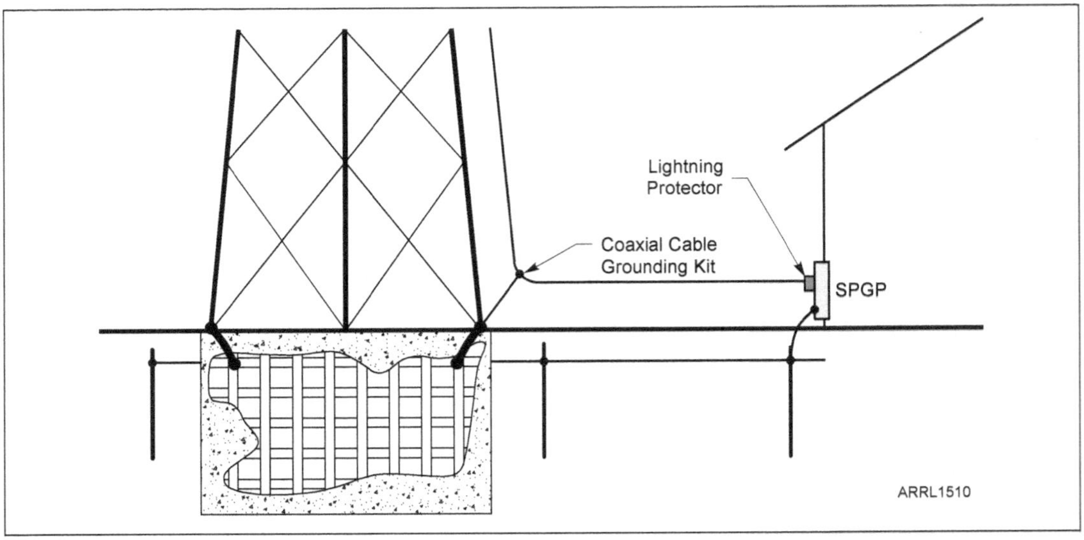

Figure 4.8 — Grounding coaxial cables should be done at both the tower top and the base, as well as at the entry point to the house or building. For minimum energy transfer from the tower, bring the coax all the way down the tower and use a cable shield grounding kit. [Drawing from *The "Grounds" for Lightning and EMP Protection*, Roger R. Block, PolyPhaser Corporation, Minden, Nevada]

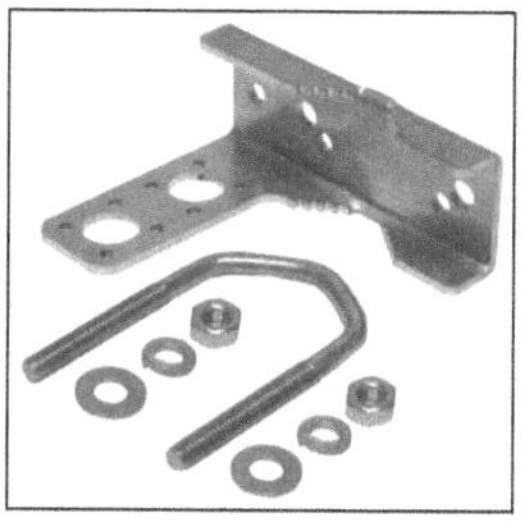

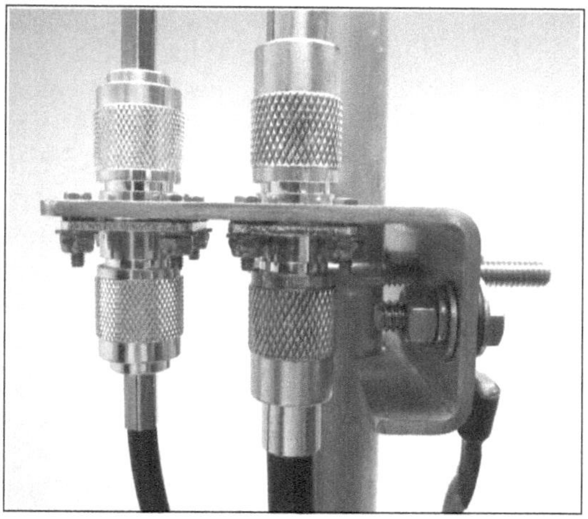

Figure 4.9 — A commercial coaxial cable grounding kit from DX Engineering (above). The same kit used to bond a pair of coaxial cables to a tower leg (right).

nector. Be sure to weatherproof the feed line connectors at the bonding points.

For towers taller than 150 feet, bonding should be done every 75 feet down the tower as measured from the top. Use the proper techniques for connecting dissimilar metals when attaching the grounding kit to the tower.

Single-Point Ground Panel (SPGP)

One of the most important parts of the lightning protection system is the *single-point ground panel* (*SPGP*). This is the point at which all antenna system feed lines, equipment power, and control lines are brought together to share a common ground connection. The SPGP must be the one and only point within the radio station that has a connection to a lightning protection ground system. In addition, protectors for other cables entering or leaving the radio station, such as telephone, rotator, TV, etc., must be mounted on the SPGP. The SPGP also satisfies the requirement of NEC 820.93 for grounding the shield of coaxial cables that enter buildings.

The SPGP serves as the mounting point for the lightning protectors associated with wiring or cables that enter the radio equipment. The purpose of each protector is to effectively short all of the conductors in an incoming cable when triggered or *over-voltaged* by a surge transient. Be aware that some protectors are not symmetrical and have a protected and unprotected side. These protectors cannot be reversed and still function correctly.

It is important that the short circuit is applied to all conductors at the same time to keep changes in voltage within the cable relatively synchronized. If the protectors are separated by more than a few feet, that can result in destructive voltage differences for short periods.

By shorting all of the conductors in the cable so they are all at the same voltage, no current can flow through the equipment between whatever is connected to those conductors. Extending this idea further, by mounting all of the protectors on a single conductive surface or panel, no current will flow between the cables or the equipment connected to them. Because the protected equipment can still be elevated to a high voltage, it would still be dangerous to be in the station when lightning is present.

If the protective devices are mounted on the same panel and connected to the same ground point, the clamping action occurs at approximately the same time. This allows voltages at the protected equipment to rise and fall together. Even if the entire system is at an elevated voltage during the transient, the difference in voltages due to timing variations is minimized. Major equipment chassis should also be bonded to the panel.

The SPGP will be different for every installation. Whatever form your SPGP takes, it must be the only common ground point for the equipment in the zone of protection you want to establish. (See the Lightning Protection Planning section of Chapter 6.)

Figure 4.10 shows a drawing of a typical SPGP. The panel itself can be copper-covered fiber board or a metal sheet. SPGP panels are available from several vendors, or you can make your own. The panel should be placed along the normal path that the coaxial cables would follow to the radio equipment. For best performance, the panel itself must be connected to a nearby ground electrode (usually a ground system or perimeter ground) with a heavy wire or preferably strap.

An ac power surge protector mounted on the panel creates a *protected line duplex outlet* or *PLDO*. The surge protector can be supplied from its own branch circuit or from an existing branch circuit. A single branch circuit can usually supply a medium-size station's transceivers and computer equipment. Amplifiers should be on their own circuit with their own ac protector. The protected circuit is created by connecting a wall ac outlet or dedicated circuit breaker to the protector on the panel. Protected ac power is then run to the station along with the feed lines and control cable and distributed to all station equipment within the protected zone.

All cables that connect to the station (or to a device that connects to the station) must have a lightning arrestor or surge protector mounted on the SPGP. This includes rotator and control cables as well as ac power. Make the SPGP large enough to separate protected and unprotected cables. Don't forget to allow for future growth of your station in the SPGP layout. Allow at least twice the space that you think you might need.

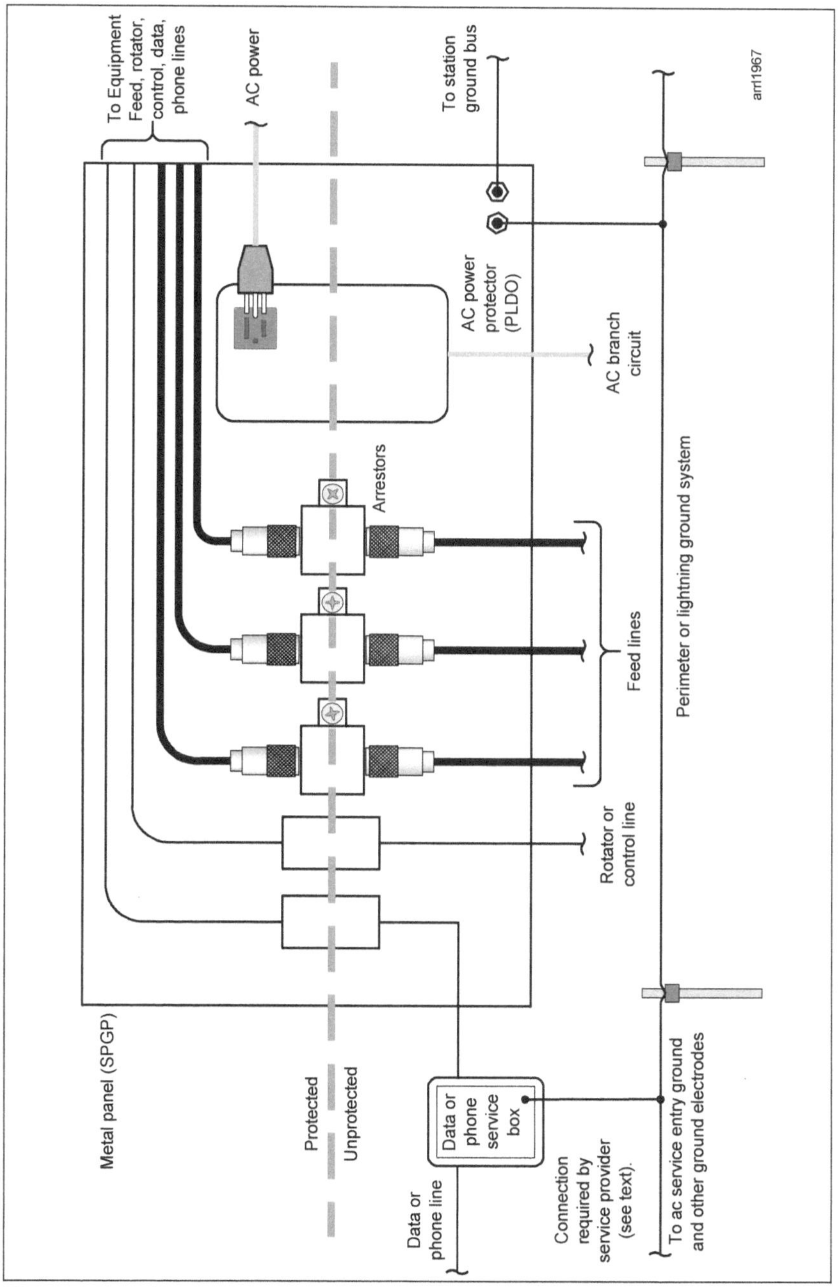

Figure 4.10 — A drawing of a single-point ground panel (SPGP) showing how various lightning protectors are mounted. The protected side of the panel is at the top in this drawing. Cables above the dashed line are protected and below are unprotected.

Telephone and data service providers will install a service box on the building where the transition is made from external to internal wiring. They are likely to follow installation rules that require grounding the box to an external ground rod. If the rod is already installed, bond it to the perimeter ground. You can also ask the installers to use your perimeter ground if available. Unless your service box is mounted on the SPGP, use a separate telephone or data line protector as shown.

Figure 4.11 is an overview of a building lightning protection system with a single SPGP. If the tower is close to the building, this approach will be sufficient for feed lines and control cables. If the tower is more than approximately 50 feet from the building, it should still be bonded to the SPGP with all feed lines bonded to the tower. Similarly, if the station equipment is more than a few feet from the SPGP, the RF bonding bus will serve as a second common ground panel. (These situations are discussed more in Chapter 6.)

Next, consideration must be given to the physical placement of the protectors on the SPGP. It is important to maintain separation between the incoming unprotected cables and the protected side of the same connec-

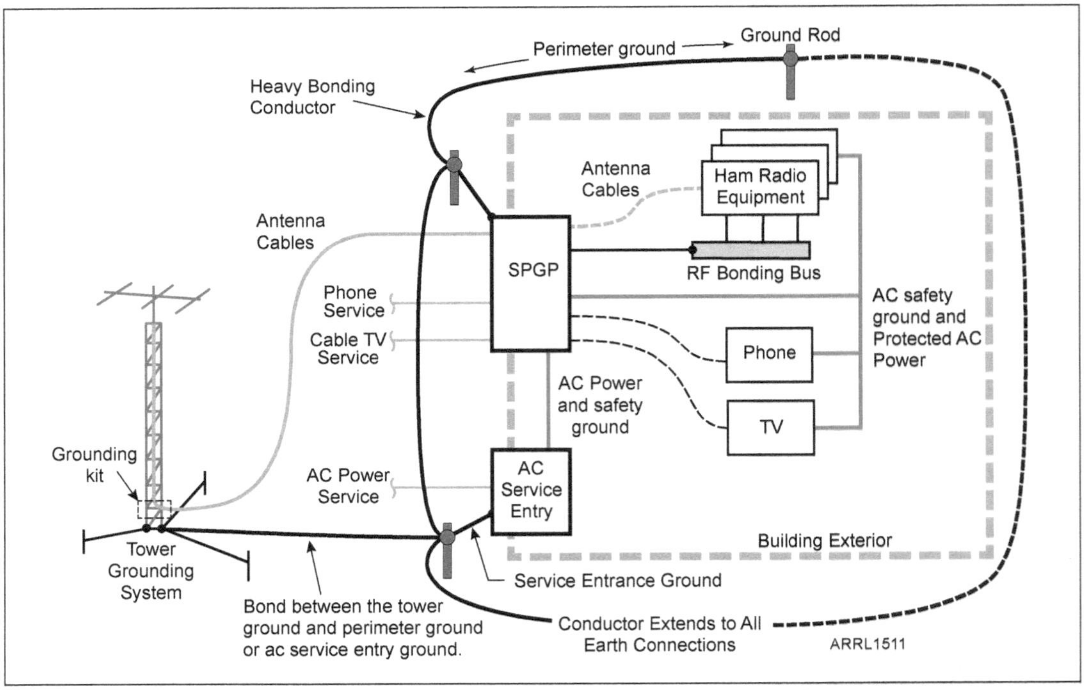

Figure 4.11 — Overview of a building lightning protection system with an SPGP. All services that cross the building perimeter are protected by an appropriate surge protector mounted on the SPGP. All ground electrodes are bonded together as well.

tions. As a result of using an in-line protector, there will be a "spark-gap level" voltage difference for a short time between the input and output sides of the protector. You must take this into consideration when planning the layout of the SPGP. (Placing and building SPGPs is discussed more in Chapter 6.)

A general guideline is to draw an imaginary line near the center of the panel as shown in Figure 4.10. Designate the area on one side of the line as protected and the area on the other side of the line as unprotected (or vice versa). Make sure you consider how the panel will be mounted, how unprotected cables will enter the unprotected area, and how the protected cables will leave the panel. If the protected area is above the unprotected area as in the figure, cables leaving the panel to the right above the dotted line must be anchored. If they are not, real world gravity will cause them to sag and come close to, maybe even touch, the unprotected cables. If this happens, during a strike event there is the possibility for the lightning energy to jump directly from cable to cable, bypassing the protector, and breaching the protected zone.

Neatness counts — all cables connected to the SPGP (feed lines, power, control, and so on) should be cut to length and routed neatly and cleanly to and from the SPGP. This applies to wiring between all pieces of equipment, as well. Take the most direct practical route.

The SPGP must be bonded to the tower ground radial system if the tower is within 50 to 100 feet. This connection should be buried and use the same material and ground rods as used for the radials. Remember that the better the tower grounding system, the lower will be the impedance in the lightning's path to the Earth. The inductance of the bonding conductor to the SPGP ground electrode will also help shunt more of the lightning's energy into the tower ground system. If no tower or other external ground system is available, consider providing extra ground rods and radials for the SPGP ground system.

In-Station Equipment Bonding

To keep all of the equipment at the same voltage, bond the equipment enclosures to a shared bonding bus. A single, low-impedance connection to the SPGP can then be made.

For small to medium size stations, where all the equipment fits on a desk or table top, a single bonding bus is usually employed as in **Figure 4.12**. A single strap or heavy wire connection to the SPGP is usually sufficient. The Good Practice Guidelines chapter includes installation examples that illustrate several ways to accomplish this. In a portable or temporary station, equipment can be bonded directly together.

For stations with freestanding cabinets or racks in addition to an oper-

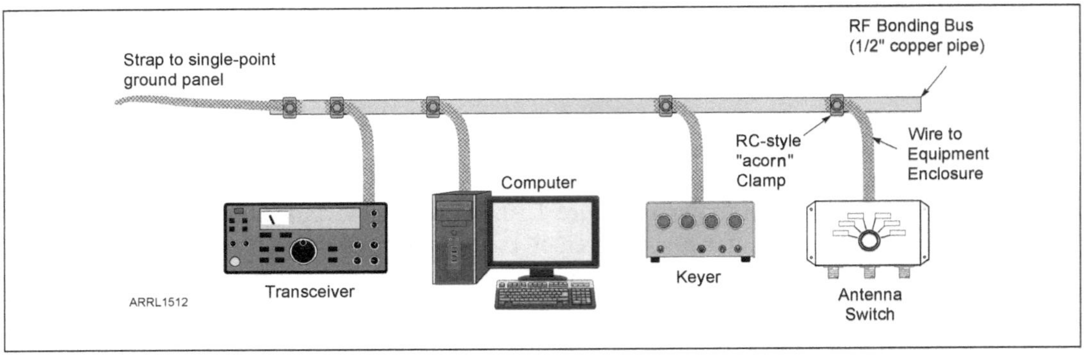

Figure 4.12 — A typical bonding bus for a station with all equipment on one table or desk. Connections to the equipment should be made with heavy wire (#14 AWG is suitable) or strap, using properly listed clamps or screws. According to the NEC, ground jumpers should curve toward the bonding bus in the direction of the ground connection although this is not practical in many amateur stations. See the chapter on good practice guidelines for a better discussion of a bonding bus and the associated connections.

ating desk, the issue of time becomes more significant due to the electrical distance between the different sets of equipment. This requires separate direct ground connections from the cabinet or rack to the SPGP. In addition, stations of this size may have other special considerations, such as concrete floor conductivity, that are not covered here. In these cases, review the commercial reference, *Standards and Guidelines for Communication Sites*, R56, from Motorola. See the chapter on Good Practice Guidelines for more information on rack-mounted equipment.

Some stations do not have access to an earth connection, such as those in apartments or on the upper floor of a residence. The most important thing to do is to create an SPGP with the appropriate protectors and to bond the equipment chassis to the SPGP. Then make a single low-impedance connection between the SPGP and the nearest suitable grounded conductor. The Good Practice Guidelines chapter discusses this type of station in more detail.

Bonding Materials

The impedance of any connection to a ground electrode should be low so the energy prefers this path and is dispersed harmlessly. Similarly, a bonding connection between pieces of equipment should be sufficiently low-impedance that even under lightning surge conditions, the voltage between pieces of equipment is very low. To achieve these goals, bonding connections need to be short (distance), straight, and wide.

Short — All conductors, no matter what size or shape, have inductance that increases with length. Connecting the SPGP to the external ground system should be done with the shortest possible connection. Short

Using Braid Instead of Strap

The standard for grounding in the communication industry is solid strap or heavy wire. Both can be used indoors or outdoors. Flat-weave, tinned grounding braid can be used if the equipment is subject to vibration or needs to be moved around. No type of braid should ever be used if it will be exposed to moisture or corrosive chemicals. Corrosion on the surface of the small wires used to make up the braid reduces its effectiveness by raising the surface resistance, where the RF currents flow. Poor contact between the individual wires can also result in noise and mixing products. Unless using braid is absolutely necessary, use solid strap or wire.

It bears repeating from the previous chapter to not reuse braid removed from coaxial cable. Once removed from its protective jacket, the braid wires immediately begin to loosen and oxidize or corrode. This reduces the braid's effectiveness at conducting RF quite a bit, making it a poor choice for long-term grounding conductors.

conductors also minimize time differences, helping keep equipment at the same voltage during a surge or transient event.

Inductance is proportional to the length of the connection between the SPGP and the ground, as well as the number and sharpness of the turns. Minimize both, regardless of what type of conductor is used. In the station, connections between pieces of equipment should be similarly direct, although some excess wire or strap is usually required in order to be able to move the equipment as required.

Straight — Rarely is it possible, in the context of an amateur radio station (unless the structure was designed around the radio station), to go directly from an SPGP to the external ground system in a short, straight line. Most of the time we are encumbered with an existing structure that is less than ideal and with esthetic constraints regarding just how much of a mess we can make. So, we do the best we can. "Straight" becomes a relative concept. Run the ground conductor as directly as possible.

Keep in mind that every time the conductor makes a turn, the inductance of the path is increased a small amount — approximately 0.15 μH for a 90-degree turn in less than 1 inch. The cumulative effect of several turns could be meaningful.

Also keep in mind that speeding electric fields don't like to change direction. The inductance in each bend or turn causes a large change in the fields over a short distance. If the change is large enough, an arc may form and we lose control over where the energy travels. Most guidelines recommend that conductors used for lightning protection connections have a minimum bend radius of approximately 8 inches.

Wide — No matter what size, conductors have inductance. Larger sizes have slightly less inductance than the smaller sizes. By the nature of its current (magnetic) fields, a wide conductor (such as strap) has lower inductance per length, compared to round conductors, and has slightly lower inductance for turns. The best choice for bonding and other ground connections is copper strap. One and a half inch wide, #26 AWG (0.0159 inch) copper strap has less inductance than #4/0 AWG wire, not to mention that it is less expensive and much easier to work with. We can use thin copper strap to conduct lightning surge energy safely because the energy pulse is of very short duration and the cross-sectional area of this strap is larger than #6 AWG wire. The strap has a large surface area that makes it ideal for conducting the strike's RF energy. This is the preferred conductor for connecting a station bonding bus and an SPGP. It is even recommended for the tower radials and the perimeter ground.

For the connection from the SPGP to a ground electrode, the goal is to make the ground path leading away from the SPGP present a lower impedance than any other path. In order to achieve this we need to find the total amount of coax surface area coming to the SPGP from the antennas. The outer shield circumference of a single RG-213 coaxial cable represents about 1 inch of effective conductor area. To make our ground path appealing to the surge energy, we ideally need more than 1 inch of effective conductive area leaving the SPGP. Where the use of a single 1½ inch wide conductor leaving the panel is reasonable, a strap 3 or more inches wide would be better. For three ⅞-inch hardline feed lines, a minimum strap width of 3 inches would be needed, and 9 to12 inches would be better.

Flat strap also exhibits the *lateral skin effect* in which current flowing on and in the strap is forced to the edges as frequency increases. At the frequencies of concern for lightning protection, this is not a great concern. At higher frequencies (at and above the upper HF range) this effect begins to limit the effectiveness of strap, and round wire becomes more effective.

4.4 Protecting Against Voltage Transients

Using effective grounding techniques helps direct current from lightning away from your equipment and into the Earth. Voltage pulses or transients will still be present, however. Currents flowing in resistance and inductance are one source. The magnetic field of currents can also induce voltages in nearby wires and cables, as well. The second line of defense, then, is to use surge protectors or transient suppressors that dissipate or divert the energy from these transients. Note that any kind of lightning protection device should not be installed near combustible material (NEC 810.20) such as wood or non-flame-retardant plastic. Further, keep flammable materials such as fuels or chemicals away from lightning protectors. The electrical discharge around these devices during a lightning strike easily has enough energy to ignite materials that will burn.

Coaxial Protectors

Coaxial protectors are unique in that they should not add to system SWR or signal loss, and at the same time they need to operate over a very broad frequency range at both receive and transmit power levels. Coaxial protectors that are used on transmit antenna feed lines must not trigger at normal transmit power, including the effects of reasonable SWR, such as 2:1. Remember to account for the correction factor of 1.41 from RMS to peak voltage when specifying the protector.

Each coaxial line leaving the station's zone of protection must have an appropriate coaxial protector. (Identifying this protected zone is part of your plan as we will discuss in the chapter on Good Practice Guidelines.) All protectors (for feed lines and other cables, as well as local power) must be mounted on a common panel such as the SPGP described earlier that is connected to an external ground system.

Two typical protectors for amateur radio use below 1 GHz are shown in **Figure 4.13**. While both of these protectors are shown with UHF-female connectors on both the antenna and equipment sides of the protector, type N connectors are available, as are combinations of male and female connectors. PolyPhaser, DX Engineering, Alpha-Delta, and other vendors manufacture or sell this type of lightning protector.

These protectors consist of a gas discharge tube (GDT) connected between the coaxial center conductor and the grounded protector housing. DC blocking protectors have a capacitor in series with the center conductor, and the GDT is connected between the antenna side of the capacitor and the protector housing.

Figure 4.14 shows how a lightning protector is installed in a feed line. Keep in mind that in order for a gas discharge protector to not break down under normal transmit conditions, the allowable firing voltage of the gas gap may well be in excess of allowable safe limits for receiver inputs. You may need to provide additional protection for the receiver input, such as a front-end protector, or disconnect the transceiver manually when not in use.

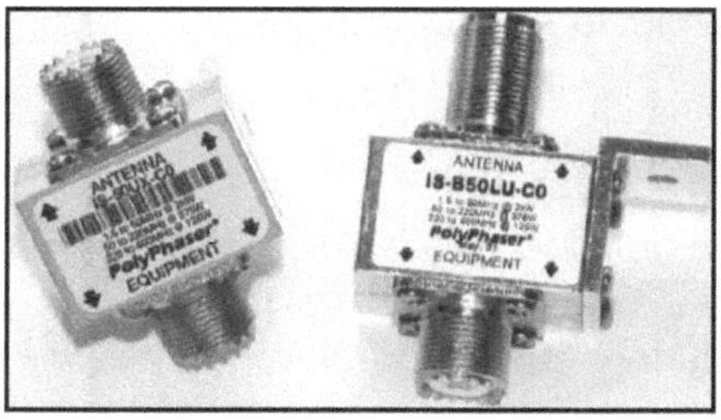

Figure 4.13 — A pair of coaxial lightning protectors manufactured by PolyPhaser — the IS-50UX and IS-B50LU.

Protectors with dc isolation have a series capacitor between the GDT connection and the protected equipment. The capacitor blocks dc and acts to slow the rising edge of the voltage transient. This makes it easier for the GDT to fire before the transient travels farther down the line toward the equipment. These types of protectors cannot be used if the feed line is to be used to supply dc or low-frequency ac voltages to a preamp or remote antenna switch.

Protectors without a series capacitor share the strike

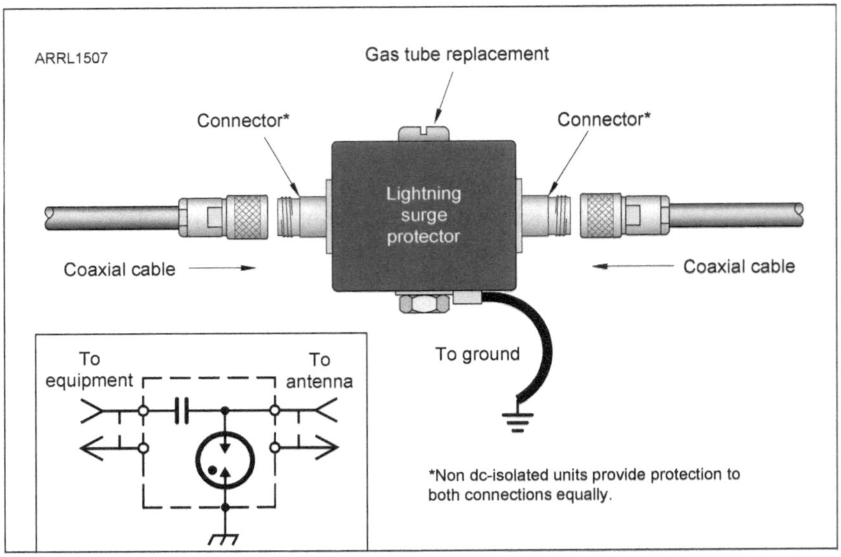

Figure 4.14 — Typical installation of a lightning protector for coaxial cable feed line. The protector is mounted on an SPGP or other grounded structure. The inset shows the schematic of a protector with dc isolation that causes the GDT to fire more quickly. (See text.)

energy with the connected equipment until the voltage exceeds the gas tube breakdown voltage. Depending on the power rating of the protector, this voltage may be several hundred volts, which can be damaging if no other protection is in place. Be sure you use the correct type of protector for your station.

Each coaxial line entering the station must have a coaxial protector on the SPGP. This also applies if you intend to use a coaxial switch, even one with built-in surge protection. The switch must be on the protected side of the cabling, since the built-in protection and mechanical contact size and spacing are no match for strike energy.

Special coaxial protectors to protect feed lines for GPS, DBS, broadcast, and cable TV are available. All protectors come with the appropriate type of connector commonly used for these applications.

Open-Wire or Ladder Line

Although protecting an open-wire or ladder line is not as convenient as with coaxial cable, some protection is warranted and possible. Commercial units are available from Array Solutions and DX Engineering that use the same type of GDT as coaxial protectors. Spark gaps as described in the next section have been used for many years by amateurs and AM broadcast stations.

You can also build your own protector by selecting two identical gas tube protectors and connecting them from each leg of the feed line to the perimeter ground near the building entrance. Each gas tube should be specified as capable of handling an instantaneous peak current of approximately 50,000 A based on the 8/20 μs IEEE standard test waveform (see Figure 4.3) and have a breakdown voltage that is well above the normal transmission line operating voltage. Typical breakdown voltages range from 600 to 7500 V, and high-surge units with ratings of 10 to 20 kA are sufficient. Be sure to consider the highest SWR and the highest transmit power in your calculations. (Remember to use peak voltage values as mentioned earlier.) The voltage chosen should be about twice the maximum calculated feed line voltage to minimize the potential for the accidental firing of the gas tubes during tune-up or transmitter anomalies. Remember that the farther the protectors are from the SPGP, the greater the timing difference with respect to other protectors on the SPGP.

Keep in mind that the application of GDTs or spark gaps to the open-wire or ladder line represents a shunt-type connection. That means that the transmission line will share a significant amount of lightning strike energy with your equipment before the gas tubes begin to conduct. Unfortunately, this type of transmission line makes it difficult to achieve a high level of confidence in protecting sensitive equipment, such as receivers, which may need to be disconnected when not in use.

The NEC also allows a switch to be used that shorts the open-wire line conductors to a ground connection (see NEC 810.57). The typical hardware store DPDT knife switch is not rated for this job. The switch and the attached grounding conductors need to be able to carry full strike current. You will need to use a heavy-duty knife switch, which is hard to find new, but some can be found via surplus dealers or used. Even when shorted, a significant amount of strike energy will be shared with your equipment. Electrical power switches such as those made by Filnor, Square D, Eaton, and other industrial manufacturers will also work but they are expensive. A less expensive and safer approach is to install a weatherproofed balun outside of the building and terminate the open wire line at the balun. Then use coaxial cable and a suitable coaxial protector as described earlier.

AC Power Protection

AC power protectors are available in many shapes, capabilities, and methods of connection. Some caution should be exercised in choosing your protector. There are many inexpensive power line protectors on the market that are clearly not suitable for lightning protection. Many of these protectors depend on the safety ground wire to carry away the surge energy. While the safety ground may provide a dc path to ground, the #14

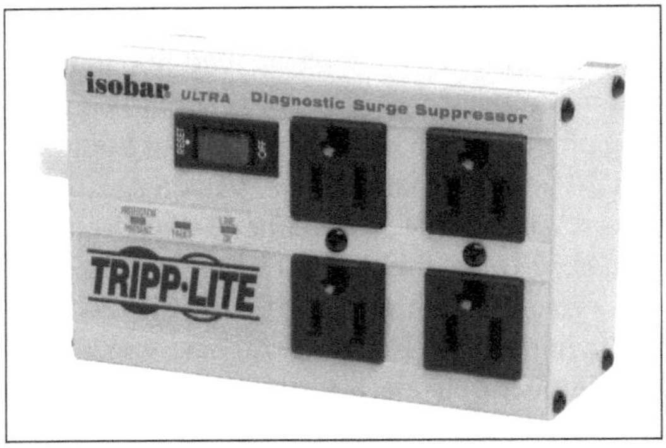

Figure 4.15 — An in-line ac power protector suitable for lightning protection.

AWG wire commonly used is too inductive for the strike currents that it must conduct to ground. The result places high voltage on the enclosures of equipment supplied through the protector. The solution is to use power protectors like that in **Figure 4.15** that can be mounted directly to the station's SPGP.

In addition, some low-end manufacturers who do provide in-line ac protectors use ferrite core inductors to maintain a small sleek physical appearance. While this approach works well when the protection is merely handling power line noise, the inductor saturates under the massive current of a real strike and the benefit of the inductance disappears. Plastic housings and printed circuit boards should be avoided where possible since they will most likely not hold up under real strike conditions when you most need the protection.

When establishing a local zone of protection for the radio room, you need to choose an in-line ac power protector matching your voltage and current requirements. For most small- to medium-size stations, a single 120 V ac protector with a capability of 15 or 20 A will satisfy all of the ac power needs. Each of the electronic items with an ac power line in the zone of protection should be aggregated into a single line as long as it is comfortably within the maximum amperage of the selected protector (usually 15 or 20 A) and electrical wiring. Larger stations with high-power amplifiers will most likely have a separate 120 V ac or 240 V ac power circuit that will require a separate ac power protector.

Surge protectors suitable for creating a PLDO as described earlier are available from companies such as Tripplite (**www.tripplite.com**) and Morgan Systems (**surgestop.com**). The Isobar model shown in Figure 4.15 is typical of shunt-mode surge protectors based on MOVs. SurgeX (**www.ametekesp.com/surgex**) makes series-mode protection models in a variety of configurations suitable for branch circuit use. (See the following section for a discussion of shunt- and series-mode protection.)

Whole-House Surge Protectors

Several companies offer "whole house" surge protection that is installed at the ac service entry panel. (Eaton, Intermatic, Leviton, Siemens, and Square D are well-known examples. Your electrician or electrical contractor can recommend devices suitable for your home.) These offer protection only against transients on the incoming ac power lines. These do not create a zone of protection for the whole house. They only remove energy from the house ac wiring system and help keep lightning transients from entering the station's zone of protection. They also do not protect against transients induced on branch circuit wiring. There must be separate protection specifically for the amateur station's zone of protection or for some other electronic device such as an entertainment system.

Protection should cover all three paths: line-to-ground, neutral-to-ground, and line-to-neutral. There should be an LED indicator that shows when protection is active in case a surge disables any of the capabilities. Some manufacturers offer models that can protect CATV and telephone lines, as well. By having all of those functions incorporated into one unit, they all share a common ground and thus all are at a uniform potential during a strike.

Telephone

Although less common with each passing year, *landline* or *plain old telephone service* (POTS) is still fairly common. This is a balanced line with a –48 V dc circuit and up to 140 V ac ringing voltage. The telephone company provides personnel safety protection in the service box where the inside *premises wiring* is attached. This service box should be grounded to your external ground system as discussed in the SPGP section. The telephone company protector is not sufficient to protect a POTS line entering a radio station. An in-line protector must be installed on the SPGP and the protected premises wiring run from that point.

There are a number of surge protectors available for telephone equipment inside a building. If you have DSL service that uses the same wiring, be sure that any surge protector is compatible with DSL signals. A word of caution — many of the protectors on the market use modular connectors (RJ11, RJ12, RJ45). This is a light-duty connector, and common amounts of surge energy are very likely to destroy the connector by welding the contacts together or fusing (melting) them open. Telephone line protectors for all typical services are available from companies such as PolyPhaser (**www.polyphaser.com**).

In some cases where bandwidth or performance speed is critical, a fiber optic connection is recommended. Similar to Wi-Fi, there is no copper connection and it therefore does not require any additional protection.

Wired Data Networks

Ethernet network cable connections must also be protected as a part of the protection plan. If there are Ethernet network connections linking the amateur station to the outside world or a computer in another room, each connection must also be protected as a part of the protection plan.

Figure 4.16 — This multi-line protector can contain up to five PC boards. Each PCB is configured for a specific type of protection. [Ron Block, NR2B, photo]

Ethernet is also commonly used to connect a broadband router to a fiber optic service box outside the building or a local coaxial-based router/switch to the computer network. For up to 100 Mbit UTP (unshielded twisted pair) networks, use protectors for CAT5 cable (four-pair). (See **Figure 4.16** for an example of a data line protector.) This protector is wired in series with the network and grounded similarly to other protectors. For higher-speed networks, use STP (shielded twisted pair) CAT6 cable and protectors rated for the higher speed data signals. Data protectors are available from companies such as PolyPhaser (**www.polyphaser.com**).

It is highly recommended that a Wi-Fi link be used instead of wired Ethernet. Wi-Fi allows data links with equipment outside the zone of protection without requiring additional protection. Wireless data also eliminates the rats-nest of cabling between devices which is very hard to protect from lightning-induced surges, and it reduces the amount of RFI to and from the Ethernet devices. Note that a wired connection extending beyond the zone of protection from a broadband gateway, router, or access point must still be protected.

Figure 4.17 — A multi-line protector for the typical 8-conductor rotator control cable. This shunt-type device is capable of protecting up to eight circuit lines with an operating voltage of up to 82 V dc.

Control Circuits

Control circuits for all external devices must be protected, especially those that are tower-mounted. In the amateur station, the most common such devices are antenna rotators and antenna switches. Since most of the control circuitry for these devices is mechanical switches (as opposed to electronic), we can use a less expensive shunt-type protection device, such as that shown in **Figure 4.17**. Control circuit protector providers include PolyPhaser and Morgan Systems.

If station ac is sent outside for convenience, for safety lighting, or to run motors (not the common antenna rotator), then that ac circuit must be separately protected as it leaves the radio room or building to prevent external energy from entering the zone of protection. The power protector(s) must be mounted on the SPGP.

Some newer rotator models use optical encoders and a modestly protected digital interface. These must also be protected. The method of protection will change, however, since the interface is electronic. Once the peak operating interface voltages are determined, it is relatively straightforward to choose the appropriate inline protector for the individual conductors.

Miscellaneous

Depending on the equipment in the radio room, there may be additional lines remaining to be covered. This section addresses a few of the more common types; the others will probably require some special attention based on the physical conditions of the site.

For those radio rooms that have external receiving antennas, protection is similar to the coaxial protectors described above, except that the impedance of the unit is 75 Ω and F-type connectors are used. For cable TV, your service provider should have lightning protectors available if they are not already installed in the service entrance. If the TV is in the station, then the protector must be on the SPGP, even though there may be a protector at the cable entrance to the building. As with telephone and data services, the cable TV service box ground should be connected to the SPGP if it is in the radio room, or else to the external perimeter ground.

For single and dual-LNB DBS dishes, the protector is required to have a very broad band-pass and pass dc through the coax center conductor.

GPS feed lines also are commonly required to carry a dc voltage. A high-quality protector will separate the RF from the dc and protect each to its own voltage and power specification.

Other types of lines and services will require special attention based on their characteristics and the physical nature of the station and residence or building. Be sure to bond their ground connection to the perimeter ground or SPGP.

4.5 Transient Suppressors and Surge Protectors

There are many manufacturers supplying protective components, so a list of types and part numbers is not provided in this book. Distributors of component-level devices typically provide access to data sheets and selector guides on their websites. Parts lists for amateur products are also helpful in identifying the type and ratings of device for a particular application. References to manufacturer application notes and product selections are provided at the end of each section. (These links are current as of February 2021.)

Series and Shunt Devices

There are two types of transient or surge protection devices — series-mode and shunt-mode — which refer to how they are connected with respect to the equipment or circuit being protected. Series-mode protectors store the surge in a very large inductor, then discharge it slowly after the strike. Due to the size of the inductors needed, they are only practical for use on branch circuits with a neutral bonded at the service panel. A series-mode protector must be able to withstand the full signal level to or from the equipment without dissipating too much energy or significantly altering the load characteristics.

A shunt-mode protector is installed between the protected conductor and a local ground system, such as a chassis connection or SPGP. The protector acts as an open circuit until a threshold voltage is exceeded, and it then changes to a low-impedance state to dissipate the transient energy as heat like a resistor or direct it to ground as a switch. Shunt-mode protectors divert strike current to the point at which they are bonded. (Series-mode protectors do not divert current to the ground system.) If equipment is interconnected between shunt-mode protectors bonded at different points, such as different outlets, destructive voltage differences can result. MOVs inside equipment also divert current and create voltage differences. This is why using a single heavy-duty surge protector is better than several smaller units.

A shunt-mode protector "shares" the transient voltage with the protected equipment until the protector's threshold voltage is exceeded. The protected device is exposed to the transient energy until the shunt device switches to the low-impedance state. Most protective devices used by amateurs are shunt-mode, such as lightning arrestors and spark gaps.

Metal Oxide Varistors (MOVs)

Metal oxide varistors (MOVs) are frequently used as shunt-mode protectors, connected from the protected line to a local chassis or ground connection. As shown in **Figure 4.18**, the varistor consists of metal oxide grains pressed together and heated (sintered) into a disc that has electrodes and wire leads attached. An epoxy coating completes the component, and it looks almost identical to a disc capacitor.

The size and composition of the grains determine how the MOV behaves electrically, as shown in Figure 4.18. At voltages below the MOVs breakdown voltage, the MOV acts like an open circuit. As a transient exceeds its breakdown voltage, the MOV begins to conduct and acts as a low-value resistor. (The voltage-current graph for a fixed resistor is also shown for reference.) This allows the MOV to dissipate the transient's energy as heat. The MOV is non-polarized and acts the same with either positive or negative voltage across it.

MOVs are available with a wide range of breakdown voltages and energy ratings. Their specified operating voltage is well below the breakdown

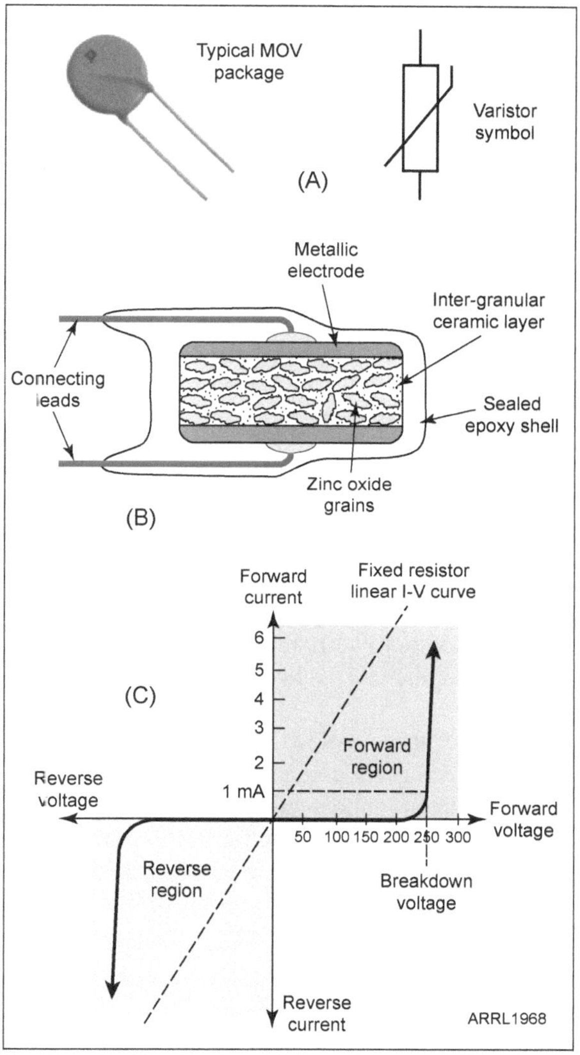

Figure 4.18 — The schematic symbol and a typical metal-oxide varistor (MOV) component (A). MOVs are made from a sintered disc of metal oxides between metal electrodes (B). The current-voltage graph in C shows the change to a low-resistance state above a threshold voltage of either polarity.

voltage, so they won't begin to conduct under normal circumstances. Their energy rating also stipulates how often the rated energy can be safely dissipated. Repeated transients cause the MOV breakdown voltage to decrease, and eventually the normal operating voltage can cause the MOV to fail at a low resistance.

The most common use for MOVs by amateurs is in ac power surge protectors and in protecting data and control lines. They are not used in high RF power environments. An MOV subjected to a very large transient may fail open-circuited so that it no longer provides any protection. Surge suppressors containing MOVs often have an indicator to show when the MOVs are no longer providing protection. Thermal fuses are included to prevent fires if the MOV fails as low-resistance.

- Bourns: **bourns.com/products/circuit-protection/varistor-products**
- Littelfuse: **www.littelfuse.com/products/varistors.aspx**
- TDK: **product.tdk.com/info/en/products/protection/voltage/lead-disk-varistor/technote/apn-disk-varistor.html**

TransZorbs

Essentially a heavy-duty Zener diode designed for transient suppression, the TransZorb (also known as a *transient voltage suppressor* or *TVS*) acts as a voltage clamp but does not dissipate the transient energy. TransZorbs are available in both unidirectional (like a regular Zener diode) and bidirectional (two Zener diodes in series) configuration. In the reverse direction, the TransZorb acts as an open circuit until its breakdown voltage is exceeded, when it begins to conduct. In the forward direction, it is a regular semiconductor diode.

The TransZorb acts faster than an MOV or GDT, both in clamping and recovery, but does not dissipate the energy of a transient. It can clamp repeated transients without the breakdown voltage changing as long as its power dissipation rating is not exceeded. TransZorbs are generally used in low-power signal applications, where more control over clamping action is required, and repeated transients are expected. Low-capacitance TransZorbs are available for high-speed data lines but are limited to low-energy clamping for static discharges and small transients.

• Vishay: **www.vishay.com/diodes/protection-tvs-esd/trans-zorb**

Gas Discharge Tubes

A gas discharge tube (GDT) is filled with a special gas and has electrodes shaped and spaced for a consistent, well-controlled *breakdown voltage* or *sparkover voltage*. (See **Figure 4.19**) This is the voltage at which the GDT "fires" (an arc forms between the electrodes), limiting the voltage in the feed line. GDTs are available with breakdown voltages of less than 100 V to several kilovolts. The GDT is connected across the protected signal or feed line so it acts as a shunt-mode device.

After the GDT fires, the voltage across the arc is limited to about 10 V, depending on the model and the current level of the arc. Once the transient energy is discharged, the arc is extinguished and the GDT returns to an open state. As long as the maximum current is not exceeded or the electrode surfaces are not damaged, the GDT will continue to provide the rated protection.

There are several types of packages as shown in the figure. The GDTs used in commercial lightning protectors for coaxial cables are removable cartridges available from the protector vendors and manufacturers. Amateur applications are considered "low to medium voltage" for rotator or antenna switch control lines and "high voltage" for feed lines. Because

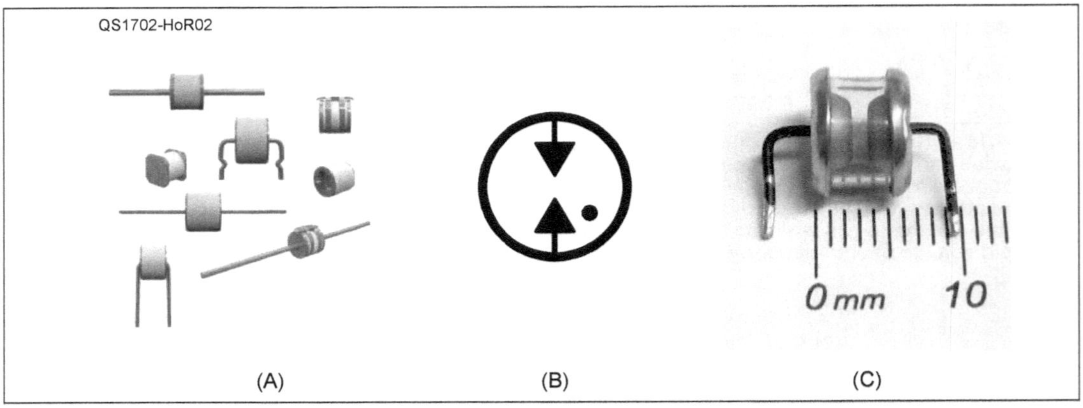

Figure 4.19 — Typical gas discharge tube components (A), the GDT schematic symbol (B), and a close-up photograph showing the electrode shape (C). [Photo provided by Ulfbastel via Wikimedia Commons]

GDTs rarely fail shorted, they are recommended for protecting rotator control lines instead of using MOVs.

Axial wire lead GDTs are also available for PC board and terminal-strip use. Testing a GDT requires a special *hi-pot* (short for "high-potential") tester that limits current through the device being tested for insulation breakdown. Motor rewinding shops often have this type of equipment and can test a GDT, but it is usually more cost-effective to replace the GDT if you think it has been damaged.

- Bourns: **www.mouser.com/pdfdocs/bourns_gdt_white_paper.pdf**
- TDK: **product.tdk.com/info/en/products/protection/voltage/arrester/technote/apn-arrester.html**
- Littelfuse: **littelfuse.com/~/media/electronics/product_catalogs/littelfuse_gdt_catalog.pdf.pdf**

Spark Gaps

Along with GDTs, a spark gap can also provide protection by limiting maximum voltages on towers or antennas or feed lines. Spark gaps are easy to make and inexpensive, too! **Figure 4.20** shows two homemade spark gaps used on insulated-base towers. Spark gaps are nothing more than metal surfaces, one connected to a ground system, that are spaced just far enough apart that normal transmit signal voltages do not cause an arc between them.

With dry air having a breakdown voltage of about 3 kV/mm, a 3 mm separation will arc over at about 9 kV. Once the arc starts, the voltage across it is very low until the current is discharged and the arc is extinguished. Be aware that the breakdown voltage may vary significantly based on the environment at the time of use, such as during rain or fog.

The spark gap in Figure 4.20A, made by Tom Rauch, W8JI, consists of a simple piece of copper pipe that is connected to a tower ground system. The pipe is positioned close to the base of the non-grounded tower section. The separation is about ⅛ inch (approximately 3 mm).

An alternate method in Figure 4.20B was constructed by the author and uses #2 AWG tinned copper wire that was left over from tower ground systems. One piece of wire is connected to a listed ground clamp on the tower leg above and below the fiberglass insulator section. The wire position and spacing can be adjusted by loosening the clamp screws or just pushing on the wire.

Another option shown in Figure 4.20C is to mount a conventional spark plug or GDT across a stand-off insulator to a grounded bracket and attach the protected conductor to its top terminal with a short flexible lead. The spark plug gap is easily adjusted to the desired width using a feeler gauge. Do not use resistor-type spark plugs.

(A)

(B)

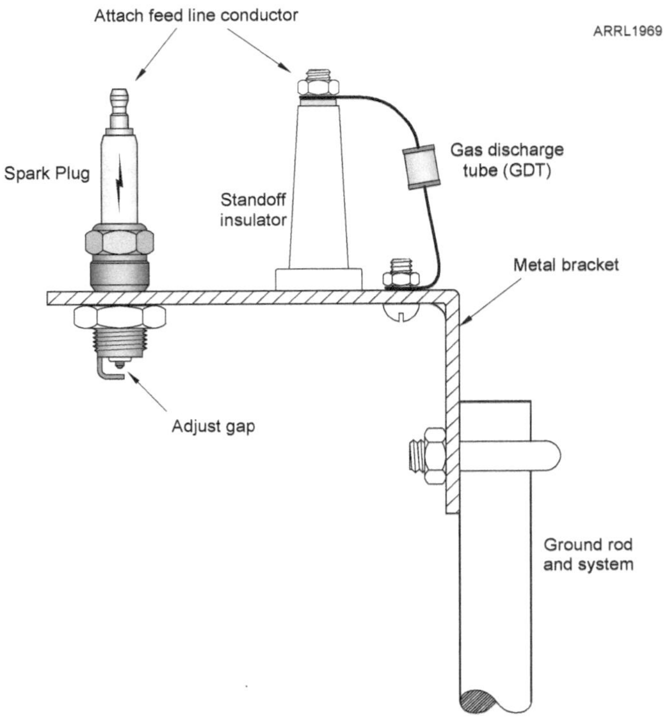

(C)

Figure 4.20 — Pipe spark gap constructed by Tom Rauch, W8JI (A) and a wire-and-clamp spark gap constructed by the author (B). Spark plugs (C) can be mounted on a grounded bracket and adjusted to the desired breakdown voltage. [Photo for 4.20A courtesy of Tom Rauch, W8JI]

Debris or water in the spark gap will reduce its breakdown voltage. Arrange the conductors so that water flows away from the gap and does not pool or gather in the gap. Keep the gap from becoming an attractive nesting site for insects. Regular inspection and cleaning of the gap is a good idea, as well. Be sure to follow proper procedures for connecting dissimilar metals, even for spark gaps.

Chapter 5

RF Management

The third aspect of grounding and bonding involves the RF energy your station creates and radiates as best it can. Being close to a transmitting antenna means that RF will be picked up by your equipment and the connections between it, including the ground system. To deal with it effectively, you'll have to look at your station in new ways.

What Do I Need to Learn from This Chapter?

- Consider the RF behavior of your whole station
- Understand the idea of "ground" from ac through RF
- Learn some RF fundamentals
- Apply bonding to equalize lightning, audio, and RF voltages
- Block RF current with ferrite common-mode chokes

Fortunately, the work you've already done for lightning protection means you are already thinking in terms of bonding and controlling current. We'll just extend those ideas to higher frequencies and group them together as "RF management," the title of this chapter.

The Nature of Things

It is important to realize that Mother Nature does not see our various types of connections as different grounds, or even as having a different purpose. Our intent in making the connection doesn't matter. There is no "sorting" of "ac power, lightning surges, control, audio, and RF between the different cables and wires." These are just wires connected to different places and current will flow according to voltage and impedance. All currents flow on all wires.

It is just as important to take into account that the wires, and the points to which they are connected, all have different properties at different frequencies. The two previous chapters have presented and explained details of low-frequency ac safety grounding and lightning protection at higher frequencies.

In this chapter, our primary concern is with managing the RF currents and voltages that are present on the connections and enclosures of our station. By doing so, we will also improve the station performance at audio frequencies. Audio is not only important for high-quality voice mode operation, but also for the increasingly popular digital modes that use low-level audio signals on both transmit and receive. Your goal should be to send and receive clean signals, free from hum and buzz noise.

As stated in the chapter on Grounding and Bonding Basics, part of the intent of this book is to help you to create one grounding-bonding scheme for your station that works for ac safety, lightning protection, audio, and RF management. We are now ready to complete the circuit.

5.1 Your Station, the RF System

In a story from the early days of radio, a person visiting a radio station is said to have exclaimed, "I don't know why they call it wireless. I've never seen so many wires in all my life!" Things have not changed much!

Because there are lots of connections in even a simple station, trying to analyze each one is not realistic. Your station is best viewed as a big system of interconnected pieces. This chapter will help you understand some basic rules about building a station so that it behaves the way you expect it to at RF while satisfying the requirements for ac safety and lightning protection.

Sooner or later, just about every ham who builds an HF station at home, in the car, at ARRL Field Day, or for portable operation experiences an *RF burn*. RF burns are caused by a person touching a point at which a high RF voltage is present when the transmitter is on. Who says you can't feel RF? Although RF burns can sting, particularly from a microphone or key, they only result in a certain wariness on the part of the burn-ee and generally don't present a hazard at power levels of 100 W and below.

Similarly, you may find that transmitting on certain bands causes some equipment to misbehave, indicator lights blink, or perhaps your transmitted audio for speech or data modes may be distorted. Maybe adding a new antenna or accessory upsets the tuning of an antenna. What's happening here?

These are exciting encounters with RF floating around your antenna system. "But wait," you exclaim. "The antenna is up in the air and connected to the antenna tuner! This has nothing to do with my antenna system!" Oh yes it does! Unless your station is built inside an RF-tight metal enclosure or otherwise isolated from the antenna and feed line, every coax shield, every enclosure, every unshielded wire — any conductor that is connected directly to the transmitter or indirectly just by being close to the antenna should be treated as part of one big antenna system, including the antenna itself. This includes the wiring of your ac safety and lightning protection ground sys-

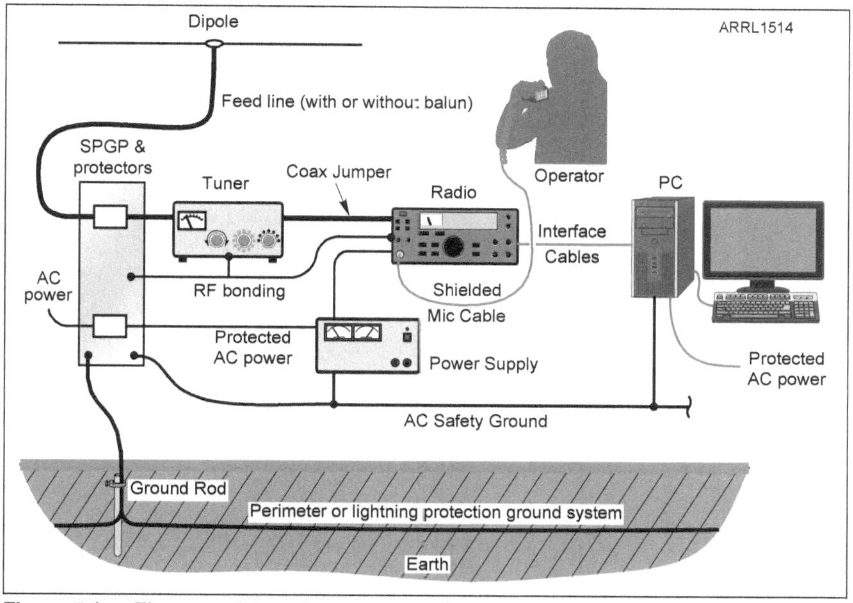

Figure 5.1 — The complete antenna system for a typical basic station. Everything conductive, including the operator holding the microphone, is part of the antenna system. (This is not a complete wiring diagram, just an illustration of connections that interact with transmitted RF. Bonding connections are largely omitted.)

tems. It also includes *you* when touching any of those conductors!

While less dramatic than "getting bit," one has to watch out for RF currents, too. RF current on the outside of an enclosure or cable shield can find a way into the electronics via an unshielded connection (such as a power cord) or an improperly connected shield. RF where it shouldn't be can wreak havoc with a circuit's operation: audio gets garbled, keyboards stop working, and control interfaces stop controlling. This is the subject of the section "Proper Shield Connections" later in this chapter.

Take a look at **Figure 5.1**, which shows a typical home station consisting of a coax-fed dipole, a transceiver and tuner, a computer connected to the radio, an SPGP connected to its lightning protection ground, the required ac safety ground, and an operator. Everything in that figure is part of the antenna system of that station! Whenever the radio is transmitting, RF will be present on each enclosure and connection in the station, including the operator!

Obviously, we would like to control the RF voltages and currents so they don't cause our equipment to malfunction or burn our fingers. The natural tendency is to think, "I'll just ground everything and it will be at zero volts — problem solved!" Not so fast! You're partly right, but we have a failure to communicate, as they say.

The Myth of RF Ground

Back when most amateur operation took place below 15 MHz, a few feet of wire connected to a ground rod was electrically short — a small fraction of a wavelength. Being short, the impedance of the wire was very low and could be ignored. If the station was on a home's ground floor or in a basement, that connection to the Earth was enough to minimize the RF voltage of equipment enclosures and accessories throughout the station. Gradually, this connection became thought of as providing a zero-voltage point at RF and the idea of an "RF ground" became entrenched.

At the higher frequencies commonly used today, that same wire running to a ground rod is electrically longer. Its impedance varies a lot, even from band to band, and often acting more like an open circuit. The wire you can treat as "0 V" at 60 Hz is a different animal at RF, which includes a lot of lightning's energy and all of our ham radio activity. On the higher HF bands and VHF/UHF, that wire can even become an effective antenna, both radiating and receiving RF signals.

This is an even bigger problem for hams with stations on an upper floor! In today's stations that can operate from past 160 meters through UHF and into the microwave realm, how can we control the RF voltages and currents in our station?

The correct approach is to stop looking for the elusive "zero voltage connection" at RF. The Earth is not a magic drain into which all of our unwanted RF can be poured through a wire. Even points on the Earth's surface separated by a few feet can look quite different at high frequencies. An electrically long connection to the Earth can behave unpredictably at RF and often causes RF-related problems on its own. So, repeat after me, "There is no RF ground!"

Nevertheless, we need to deal with the RF voltages and currents that appear in our station whenever we transmit. We want to prevent the RF from disrupting normal operation and it would also be nice not to have to worry about encountering high RF voltages. How can we deal with that as simply as possible while also satisfying our needs for ac safety and lightning protection?

5.2 RF Basics

Every connection in your station behaves differently depending on frequency, connectors used, wire size and length, and type of cable. By understanding how conductors act at different frequencies, you can choose and arrange them to achieve the different purposes you intend.

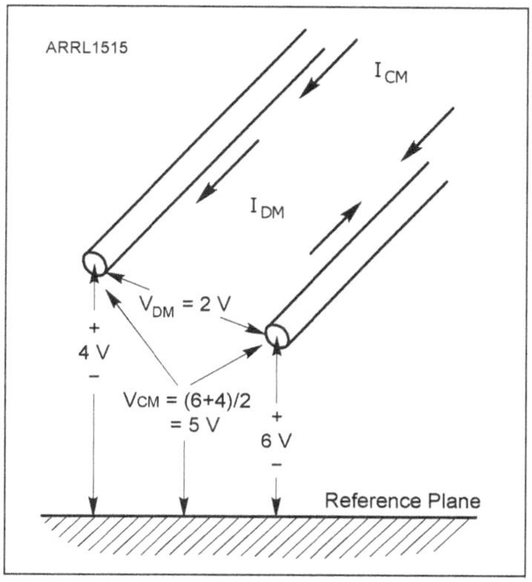

Figure 5.2 — Differential mode and common mode voltages on open-wire line are illustrated as V_{DM} and V_{CM}. Differential-mode currents, I_{DM}, flow in opposite directions, while common-mode currents, I_{CM}, flow in the same direction on both conductors.

Common- and Differential-Mode Voltages

As we proceed in our discussion, you will need to understand the difference between *differential-mode* and *common-mode* voltages and currents.

Applied to a voltage, *differential* means that the voltage is measured between two arbitrary points — a voltage difference — and not with respect to some absolute reference, such as the Earth. Differential-mode current is really a pair of currents with the same amplitudes but flowing in parallel and opposite directions, completely independent of any return path. As shown in **Figure 5.2**, a differential-mode signal exists as the voltage between two conductors, V_{DM}, and a pair of equal and opposite currents, I_{DM}, flowing on them. Neither conductor needs to be connected in any way to a voltage reference, called the *reference plane*. The reference plane can be a circuit-board ground, an enclosure, or the Earth.

A common-mode voltage, V_{CM}, has the same value on both conductors with respect to the reference plane. Similarly, common-mode currents, I_{CM}, would flow on all conductors of the line in the same direction with the same value.

A special case of differential-mode signals occurs if the conductors are very close together electrically and are either parallel to each other or concentric, forming a transmission line. The differential-mode signal is a voltage between the conductors and current flowing in opposite directions on the two conductors. Coaxial cables and open wire line are examples of tightly coupled conductors forming a two-wire transmission line. A two-wire line is a special case of a multiwire cable, such as three-wire ac power wiring, an antenna rotator control cable, or a ribbon cable for a parallel data interface. RF flowing on one of the wires will capacitively couple to the others, as well.

Common-mode voltages and currents in a ham station are often the result of conductors picking up RF. Even without RF, they can be created by differences in impedance from the conductors to the reference plane. This unbalances the conductors so that the differential-mode voltages and currents are no longer exactly equal. Common-mode is also the name for current flowing on the outside of a cable shield acting as a single conductor. (See the section on “Coaxial and Twisted-Pair Cable” below.)

Common- and differential-mode voltages and current may coexist on the same conductors but are measured differently. Figure 5.2 shows an example — imagine that a voltmeter shows one wire of a two-wire feed line at 6 V with respect to ground and the other wire at 4 V with respect to ground. The differential-mode voltage is the difference of 6 – 4 = 2 V. Common-mode voltage is the average of the two voltages: (6 + 4) / 2 = 5 V. If V_{DM} were zero, V_{CM} would be the same on both wires: 5 V.

RF common-mode current picked up from a transmitted signal is almost always present in an amateur station. It can flow on both conductors of a speaker cable or on the outside of a coaxial cable shield or on shielded audio or control cables. It helps to realize that both of the parallel conductors couple to external RF fields, and the voltages and currents created by the field will then be distributed between the conductors. Many RF-related problems can be solved by keeping common-mode RF out of equipment and out of differential-mode signal paths. Expect RF voltages and currents to be present and plan accordingly!

Shielded cable's *transfer impedance* is a measure of how current on the outside of a shield develops a voltage on the inside of the shield. This mechanism converts a common-mode current into a differential-mode voltage, which adds to the original differential-mode signal and can result in RF interference to the connected equipment. That is another reason it is beneficial to reduce common-mode current.

Proper Shield Connections

Material in the following two sections is adapted from Chapter 1 of "A Ham's Guide to RFI, Ferrites, Baluns, and Audio Interfacing," by Jim Brown, K9YC. (See the sidebar "In-Depth Tutorials and References" and the website **pin1problem.com**.)

Improper shield connections are the primary coupling mechanism for hum and buzz, as well as the primary coupling for RFI into and out of equipment. In fact, most of the RFI problems we attribute to "RF in the shack" or

In-Depth Tutorials and References

For an in-depth discussion of these and related subjects, Jim Brown, K9YC's tutorials and presentations at **k9yc.com/publish.htm** are quite detailed and we are indebted to Jim for his contributions in this area. His publications that are most applicable to this book are:

- "A Ham's Guide to RFI, Ferrites, Baluns, and Audio Interfacing"
- "Power, Grounding, Bonding, and Audio for Ham Radio — Safety, Hum, Buzz, and RFI"
- "Transmitting Chokes for VHF and UHF"
- "A New Choke Cookbook for the 160–10M Bands"

The ARRL Handbook chapters on RF Techniques and RF Interference provide additional information on closely-related topics you'll find of interest, too. *The ARRL RFI Book*, covers many techniques, especially troubleshooting, that you'll find invaluable as you build and operate your stations.

A website with a lot of good information on grounding, lightning protection, and RF interference is maintained by Tom Rauch, W8JI, at **www.w8ji.com/rfi_rf_grounding.htm**. Tom's large station on a hilltop in Georgia is a great place to learn about lightning protection techniques that work.

in audio and video equipment have improper shield connections as the root cause. The problem is also common in computer sound cards and digital interfaces. Commercially manufactured amateur equipment (except for transmitter outputs) and consumer electronics very frequently exhibit this problem at audio, video, RF, and control interfaces.

Figure 5.3 shows how coaxial cable and paired-conductor cable

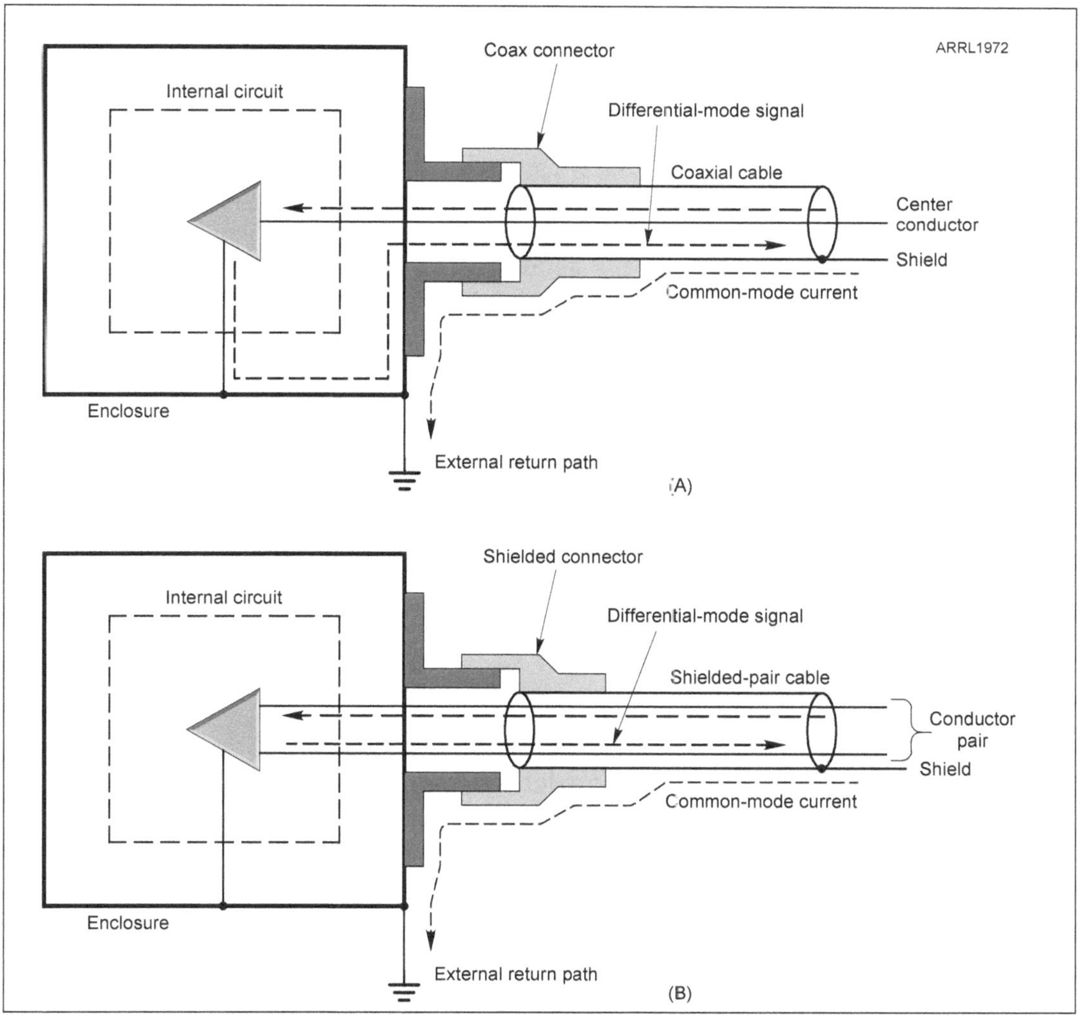

Figure 5.3 — A common-mode signal on the outside of a coaxial cable (A) is prevented from coupling to internal circuitry by connecting the shield to the metal enclosure. The common-mode signal is directed to the station's bonding conductors including earth connections. The same effects are shown for shielded-pair cable (B). The connection of the circuit and enclosure only illustrates the general current path — see the text.

shields should be connected to shielded enclosures. Multi-conductor cables should be connected in the same way as the paired-conductor example, with the signal return kept separate from the shield. (Shielding for non-metallic enclosures or with no enclosure at all are discussed in the K9YC tutorial.) Think of a cable shield as an "RF water pipe" to keep the signals you want (differential-mode audio and RF) inside the cable and the signals you don't want (common-mode RF and noise) out. (The figure is oversimplified to illustrate the common-mode current path. The actual differential signal path should form a transmission line from the cable to the electronics, such as by using an uninterrupted circuit ground PCB layer.)

Connecting the shield to the equipment enclosure keeps the common-mode current from entering the equipment, so that it flows instead on the attached bonding and ground conductors. If a shield is connected directly to a point inside the enclosure, any common-mode current will also be conducted directly to that circuit's ground reference along with the desired differential-mode signal. (See K9YC's RFI tutorial for more details.)

Part of the reason why the shield connection problem is so widespread is equipment construction. It is common for connectors such as phono jacks, tip-ring-sleeve (TRS) jacks, data connectors, and even some RF connectors to be mounted on a PC board, then extended through the enclosure. If the shields of the connectors are not well-bonded to the enclosure, common-mode current on the shield will flow directly to the internal circuitry.

If fixing the cause of the problem by connecting the shields properly is not practical, there are two common methods to use instead. The first is to bond all of the equipment together to minimize the voltage differences that cause current to flow on the shields. The second is to use ferrite chokes to block the common-mode current. These methods are described below in the sections "Bonding to Equalize Audio and RF Voltages" and "Blocking RF Current with Chokes."

Coaxial and Twisted-Pair Cable

You are probably familiar with coaxial cable, but you may be unaware of how it prevents external fields from affecting the differential-mode signals inside. Because of *skin effect*, common-mode RF current on a shielded cable will flow in a very thin layer at the outer surface of the shield, separate from whatever is going on inside the cable. At frequencies for which the shield is much thicker than the skin depth, electrical fields do not penetrate the shield and couple to the conductors.

The non-magnetic shield is ineffective against magnetic fields, however. A coaxial line rejects magnetic fields because of mutual coupling between the conductors that causes the resulting voltages and currents to cancel. For frequencies at and below the kHz range (below the *shield cut-off frequency*), however, there is little coupling between the conductors. As a result, magnetic fields can couple to the conductors, inducing noise voltages.

Twisting a pair of conductors, however, balances the noise signals picked up by each conductor so that the voltage and current in one twist cancel those in the next twist. If the twist *pitch* (number of twists per unit of length) is short and the cable construction is uniform, the cable can reject magnetic and electromagnetic field pickup throughout the HF range and into VHF. (See the sidebar, "Do The Twist.")

At power and audio frequencies, an unshielded twisted-pair (UTP) rejects magnetic fields better than a single shielded conductor (coaxial cable) and is effective against picking up electromagnetic fields well into the VHF range. This is why telephone systems and high-speed data networks such as Ethernet can use unshielded twisted-pair cables with little interference. A high-quality shielded twisted-pair (STP) cable will also reject electric field coupling over a very wide frequency range, assuming the shield is connected properly as described in the preceding section.

Peaks and Nulls

If the station in Figure 5.1 is operating on 10 meters, at least one potential hot spot is never more than about 8 feet away. Why? Consider the wavelength at 28 MHz: 33.4 feet, and so ¼ wavelength is approximately 8.3 feet. When a conductor is excited by RF, either directly from a signal source or by picking up radiated energy, a pattern of peaks and nulls for

Do the Twist

Although RF is picked up by both wires of parallel-wire cables or "zip cord," often used for power and speaker connections, the resulting induced currents and voltages are imbalanced due to unequal proximity to the source and to other conductors. Paralleling the wires forms a long, thin pickup loop, as well. This creates a differential-mode signal in the cable that can cause interference. For these reasons, untwisted zip cord offers almost no rejection of external fields.

RF rejection by parallel-wire cables can be greatly improved by replacing the zip cord with tightly twisted wires. This balances the conductors to prevent external signals from creating a differential-mode signal in the cable.

You can make your own cable from stranded THHN wire. #12 or #14 AWG works well for speaker cables. Heavier wire up to #10 AWG may be needed for dc power cables. Twist the wires with a hand drill or adjustable-speed drill at slow speed so that the resulting cable has a dozen or more turns per foot.

To maintain polarity of the cable, use two different colors of wire or identify one with a bit of tape. For ac power wiring, twisted three-conductor cable provides effective rejection. Twisting will shorten the cable by a few percent so be sure to account for that or make plenty of extra cable. Newly-twisted cable will tend to unwind a bit so let it "rest" before use.

both voltage and current is created. Peaks are ½-wavelength apart and so are the nulls, with peaks and nulls offset ¼-wavelength apart. **Figure 5.4** shows the typical pattern on a resonant wire that is three ½ wavelengths long.

Whether the conductor happens to be a wire, the outer surface of a coax shield, an equipment enclosure, or a "ground" wire makes no difference. It's all a conducting surface as far as the RF is concerned, regardless of what we call it! No matter where you are in the jumble of conductors making up the typical ham station, there can be a voltage peak within 8 feet on 10 meters.

In Figure 5.4, you can see that the peaks and nulls of voltage and current don't occur at the same point along the conductor. Where there is a voltage peak, the current is low and vice-versa. Because impedance is the ratio of voltage to current, a high-voltage point is also a point of high RF impedance and again, vice-versa. The wire in the figure is resonant because it is a multiple of ½ wavelength (½ λ, or λ/2).

Conductors don't have to be resonant for there to be peaks and nulls along the line. If the frequency changes, if the conductor's electrical length is changed, or if a choke is placed in the RF current's path, the position of the peaks, nulls, and high impedance points will also change. Any open end on a conductor creates a high-voltage/low-current point. This forces the pattern of peaks and nulls to continue all the way along the feed line and into the station unless blocked by chokes or other methods of decoupling.

Even though current reverses or voltage polarity reverses every half-wavelength, the ratio of voltage to current still repeats its high-to-low pattern. That is, impedance repeats every ½ wavelength along a conductor. Points of high and low impedance are ¼-wavelength apart. This knowl-

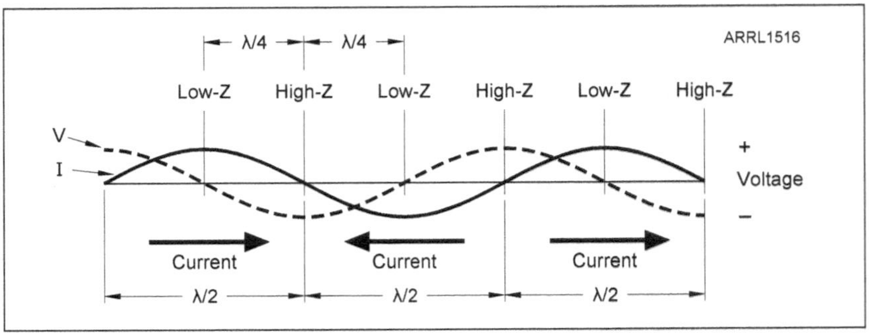

Figure 5.4 — A resonant conductor three ½ wavelengths (3 λ/2) long with a typical pattern of voltage and current peaks. Peaks of either voltage or current are ½ wavelength (λ/2) apart and there is ¼ wavelength (λ/4) between a voltage and current peak (or null).

edge can be handy when troubleshooting RF-related problems. You can learn more about variations in impedance along a conductor by studying how transmission lines work in *The ARRL Handbook* or *ARRL Antenna Book*.

5.3 Bonding to Equalize Audio and RF Voltage

The most important step in reducing audio noise and managing RF is equalizing audio and RF voltages throughout your station as much as possible. This works just like equalizing voltage for lightning protection: By bonding equipment together with low-impedance connections to make the voltage difference as small as possible, unwanted low-frequency audio noise current between the equipment is minimized and tends to flow on the bonding conductors rather than signal cable shields. A comprehensive bonding effort pays benefits at all frequencies.

It is important to remember that bonding will not result in there being *zero* audio and RF voltage on the equipment. Bonding keeps all of the equipment at more nearly the *same* voltage, minimizing the voltage differences that would otherwise cause noise and common-mode current flow between pieces of equipment.

Common-Mode RF Current

Many inside-the-station RF problems are actually caused by the way feed lines and antennas are constructed and installed outside the station! A coaxial feed line that is "hot" with RF common-mode current can create numerous issues by conducting RF right into the station where it is much more difficult to manage. See the section "Proper Shield Connections" earlier in this chapter for more about this problem.

Common-mode RF current can radiate a signal, while the same conductor can pick up common-mode current as noise. Use common-mode chokes at the antenna and on long feed lines to block this common-mode current path. A single-point ground panel as discussed in the chapter on lightning protection will also help divert RF current away from station equipment. By applying those techniques outside your station, you will have fewer RF problems inside your station.

At RF, bonding is only partially taken care of by the shields of coaxial cables between equipment. As we all know, the coax jumpers we use to connect equipment together can also be a lot longer than are necessary just to connect the equipment together. This enables them to pick up quite a bit of RF. In addition, accessories and computers and power supplies and audio interfaces generally aren't connected together with coax, so we have to provide additional bonding to minimize RF voltage differences.

Audio Frequency Noise

Many audio noise problems such as hum and buzz in low-level signals are caused by leakage currents from power wiring. (This and related topics are addressed in detail in K9YC's tutorial referenced earlier in the sidebar, "In-Depth Tutorials and References.") These leakage currents are the result of capacitive coupling between the ac line conductor and the equipment ground, mostly in power transformers and power line filters.

Audio voltages are also developed by current flowing through the resistance of the green wire (ac safety ground conductor) connecting an outlet to the circuit breaker panel. (This voltage is referred to as an "IR drop" for $V = I \times R$.) Large neutral currents from various imbalances (usually not a problem in single-family residences) can also result in different voltages at the ac outlet.

The IR drop is different on equipment plugged into different outlets, and/or bonded to ground at different points. Because of the widespread use of 3-phase distribution of ac power, that leakage current is dominated by *triplen harmonics* of 60 Hz (multiples of the 3rd harmonic such as 180, 360, 540, 720 Hz, and so on) and is heard as "buzz" rather than hum. (Pure 60 Hz hum is usually the result of magnetic fields from transformers that are picked up by loops in cables and wires.) Powering all equipment from a single outlet means there can be no audio voltage created by ground conductor IR drops between outlets or branch circuits. Minimizing audio noise such as buzz and hum is discussed further in the Good Practice Guidelines chapter.

Hum and Buzz

Hum is 50 or 60 Hz, nearly always magnetically coupled, from leakage fields from big power transformers, or fields created by multiple neutral-to-ground bonds or other wiring errors in a power system. (Most wiring errors at outlets can be found with a good outlet checker.) The strength of magnetic fields from a source, and the sensitivity of a circuit to receiving a field, are both proportional to the loop area of their respective current paths. This is why it is good practice to bundle bonding conductors with signal conductors in small-signal circuits or to run them in close proximity. Orienting the magnetic field and the affected cable at right angles can also minimize coupling.

Buzz consists of the *triplen harmonics* of ac mains power (see the text). While these harmonics are the products of 3-phase power systems and most of us receive single-phase power, much of the power distribution in residential and rural areas is a special configuration called "high-leg Delta." In this configuration, the lines on the power pole are 3-phase and feed relatively small 3-phase customers (such as farms or small manufacturing) directly in a delta configuration which does not have a ground. Single-phase customers are fed by the center-tapped secondary of a transformer whose primary is driven by one phase of the delta. (See the chapter on AC Power.) Single-phase residential customers see those triplen harmonics as neutral current, which is connected to the ac safety ground in our ac service panel. See K9YC's tutorial on power and grounding referenced in the previous sidebar, "In-Depth Tutorials and References."

Daisy-Chain and Star Connections

At audio and RF frequencies, bonding equipment directly together with heavy wire or strap, known as a *daisy-chain connection*, provides a low-impedance path for current that would otherwise flow on cable shields. This is not a recommended technique for lightning protection, however, since any voltage or current surge would arrive at each piece of equipment at different times. This can result in destructive voltage differences. (See the section "Portable Stations" in the Good Practice Guidelines chapter for more about daisy-chain connections.)

In a *star connection*, all bonding connections are directly from the equipment to the common connection point. (See **Figure 5.5A**.) Whether or not you should use a star connection depends on what you are trying to accomplish with the connection. The star connection is sufficient for ac safety and dc power grounding, where resistance is the primary consideration. The connections in Figure 5.5 are on the operating desk or table, not to some more distant point.

At the higher frequencies involved for lightning protection, inductance becomes the most important characteristic. A star configuration with short connections will provide adequate bonding, but the longer the connections, the higher the inductance. More inductance will mean higher

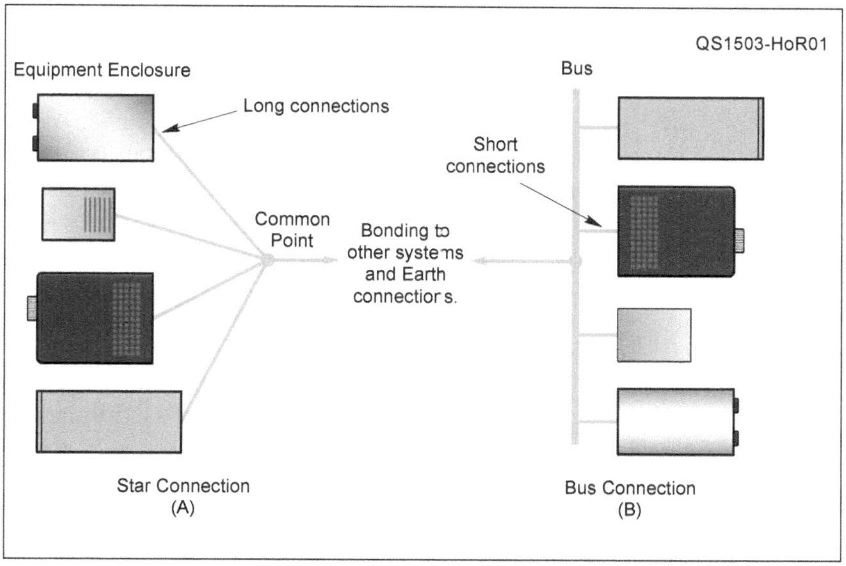

Figure 5.5 —The star connection at (A) and the bus connection at (B) accomplish the same safety functions for dc and ac power. For RF bonding, and to minimize ac power system hum and buzz, minimizing connection length and loop area makes a compact bus configuration preferable.

voltage differences between pieces of equipment. If the connections have different lengths, the voltage peaks will occur at different times, creating voltage differences between pieces of equipment. As frequency increases, the electrical length of each connection in the star increases, too. This leads to unpredictable behavior for RF bonding.

The Bonding Bus

How can a single type of inter-connection satisfy all three needs — ac, lightning, and RF? The key is short connections.

Where direct connections between equipment are impractical and connections to a star-type ground are too long, a bus does the job with short connections. To get the best results for audio, the equipment should be physically laid out in line with the signal path for audio and control lines.

A practical and effective alternative to direct equipment-to-equipment bonding or star-type connections is to provide a wide and flat or large and round conductor at the back of the equipment. (See Figure 4.12 in the previous chapter.) This creates a *bonding bus*. The bus can be any large conductor. A short length of copper pipe is inexpensive and a popular choice although metal angle or channel, or large wire or strap can also be used. You can sometimes find surplus *bus bars* in scrap metal yards, too. Connection points can be pre-drilled in the bus and left open or a screw can be installed for later use. Bonding equipment in portable and mobile stations is discussed in Chapter 6. The sidebar "Foiling RF Voltage" presents an inexpensive method of creating a temporary ground plane.

The enclosure of each piece of equipment, including computers and other non-radio electronics, is then connected to the bus with a *bonding jumper* — short, heavy wire or strap. Stranded wire works well and is flexible enough to allow the equipment to be moved or repositioned. Use crimp terminals or ground clamps to connect the bonding jumpers to the bus. It should be noted that while more convenient, use of a bus makes the bonding path longer and somewhat less effective at reducing power line frequency noise.

To satisfy the requirement for bonding all ground connections together, add a connection to the ac safety ground from the RF bonding bus. If there is a listed ac power protector mounted on the SPGP, this is already done. The bus must be connected to any separate lightning protection ground conductor with heavy wire or strap, as well. The most common situation is for the RF bonding bus and lightning ground to use the same large conductor. (When used in a discussion of ac power devices, "listed" means "certified by an approved agency," such as UL.)

Mounting equipment in a rack cabinet also bonds the equipment en-

Foiling RF Voltage

In setting up a temporary or portable station, it is rarely practical to attempt a full-blown RF bonding bus or RF ground plane. In these cases, a handful of short wire or strap jumpers can work wonders. The key is to be flexible! The author has even used a sheet of aluminum foil under the equipment with clip leads connecting the foil to each enclosure. This worked well enough to keep the RF gremlins out of a full-power portable station for a weekend! Using disposable foil is a good Field Day trick, too.

closures together. Remember to remove any paint from the mounting ears or chassis where the enclosure is attached to the rack frame. (See the chapter on Good Practice Guidelines for information on rack-mounted equipment.) Most racks have convenient terminals or mounting holes for attaching ground conductors so the whole cabinet can be connected into a ground system for ac safety, lightning protection, and RF bonding. If the cabinet is combined with equipment in a tabletop or desktop station, use wide strap or heavy wire to connect the bonding systems together. If the rack is moveable, it is acceptable to use flat-weave tinned braid as the bonding jumper to the station ground system and to any other equipment. If the rack includes audio gear integral to the station, the bond should be directly to the transceiver and as short as practical.

RF Ground Plane

An inexpensive technique that improves on the single-conductor bus is to install a flat sheet of metal under or in back of the equipment. This is often referred to as an *RF ground plane* or *RF reference plane* because the metal is connected to the ground system for the station. As discussed previously, the metal surface is unlikely to be at "ground potential" but it does create a common voltage reference, even at fairly high frequencies. The goal of an RF ground plane is to provide a lot of highly conductive area under equipment and cables so that RF voltage differences are kept small over the whole area — the RF ground plane "short circuits" the RF fields.

An RF ground plane in your station used for bonding should not be confused with the ground plane of a vertical antenna. The two may be connected together through feed line shields but are usually a long distance apart, electrically.

Figure 5.6 shows an RF ground plane in the author's HF station. The ground plane is made of aluminum flashing with a bonding bus attached for convenient attachment of bonding jumpers or straps. A piece of aluminum angle or channel in place of the pipe would work just as well. This is by no means the only (or even preferred) way to create the bus/plane combination, but it is convenient for a tabletop layout. In the photo, the copper pipe bus is held to the aluminum with pipe clamps, and the aluminum flashing is held to the plastic tabletop with screws. (Double-sided tape can also be used.) Since this is a dry environment, no corrosion prevention steps were considered necessary. If your installation will be outside or anywhere exposed to

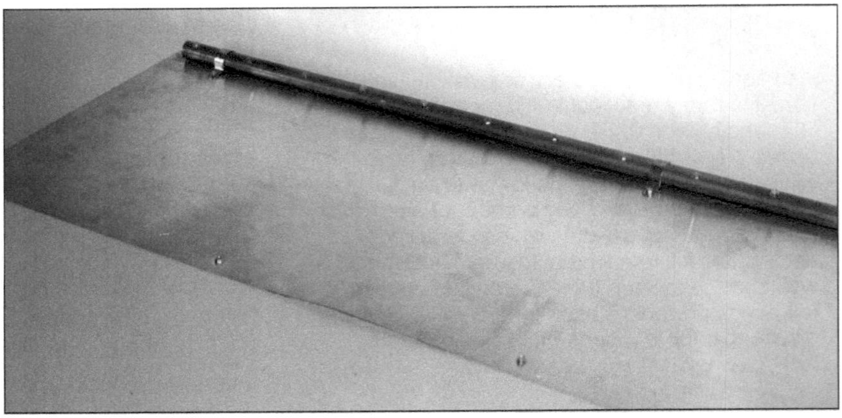

Figure 5.6 — An RF ground plane made of aluminum flashing is attached directly to a plastic table top with self-tapping screws. An RF bonding bus made of copper pipe is attached to the flashing with pipe clamps. Sheet metal screws in the pipe are available for bonding jumpers to be attached.

moisture, minimize connections between dissimilar metals or protect the connection with anti-corrosion compound.

Using a large conductive surface also helps reduce problems with audio and computer interface cables. While using just enough cable to make a connection minimizes RF pickup, commercial cables are only available in certain sizes with shortest usually 3 feet long. It's not unusual for there to be several such cables in a station with a computer or other accessories along with the transceiver. Even coiled up, these cables can pick up a lot of RF. Coiling cables also increases the amount of energy they will pick up from the large currents of a lightning strike, either direct or nearby. That's why it is important to use the shortest cable available for each connection. If the cable is coiled, make the enclosed area as small as possible. Coiling the cable in a figure-eight pattern helps cancel lightning pulse energy picked up by alternate turns.

Low-level audio signals such as for microphones or audio data signals from a sound card can easily be disrupted by RF voltages and currents. This is often referred to as *RF feedback* but usually it is RF interference (RFI) resulting from common-mode current as discussed earlier. Computer data interfaces such as USB, as well as keyboards and mice or trackballs, can also be disrupted by RF, causing the interface to malfunction or produce bad data. By placing the cables directly on the RF ground plane, they are less able to pick up RF. **Figure 5.7** shows an example of audio and data cables lying on such a metal surface.

Aluminum or copper roof flashing works particularly well as an RF ground plane, as does even galvanized sheet metal. Metal screen or mesh

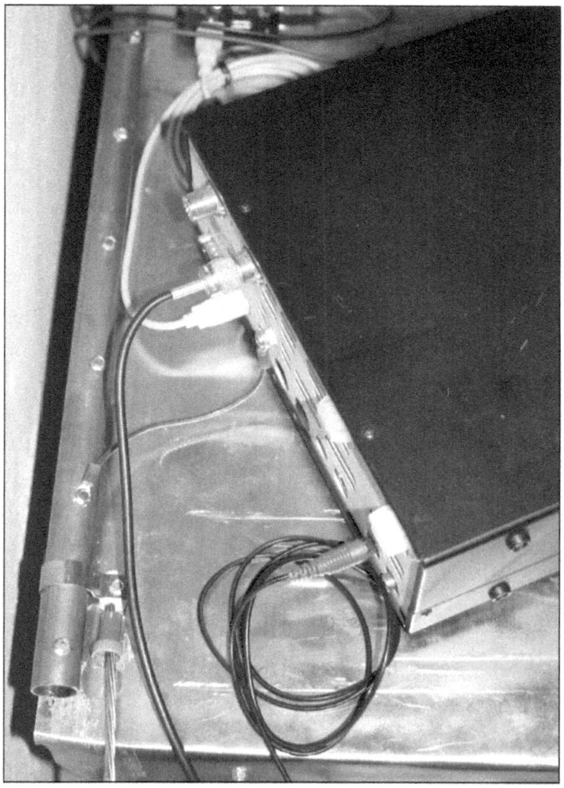

Figure 5.7 — Coiling up extra cable length and laying the cable directly on an RF ground plane minimizes the amount of RF picked up by the cable.

are good choices. A metal desk will also do nicely. Whatever you use for the RF ground plane, the RF bonding bus should be attached directly to it. If the plane material is heavy enough, it can be used as the RF bonding bus. For portable and temporary stations, baking sheets and large metal trays are excellent and are strong enough to provide mechanical stability.

At VHF/UHF/microwave frequencies, where wavelengths become quite short, the RF ground plane is the best choice for RF bonding. A bus or strap is simply too long to be effective. In most cases the capacitance between the equipment and ground plane will be sufficient to equalize RF voltages. Equipment connected to the metal surface as directly as possible will provide lower-frequency bonding for lightning protection.

Figure 5.8 is a photograph showing a station computer's interface USB cables and a USB hub sitting directly on the RF ground plane. Most computer accessories do not have a metal enclosure or a means of connecting them to a station ground system. In such cases, placing the accessory and connecting cables directly on the RF ground plane is about the best option available.

Ground Loops and Bonding

The typical amateur station is full of loops created by the shields of interconnecting cables, antenna system and control cables, ground connections, and so on. **Figure 5.9A** shows the basic idea and there are many more loops than the ones indicated in the drawing! Every complete conductive path around enclosures and cables counts as a loop. As loop area increases, so will the voltage induced in the loop, whether by a magnetic field or a rapidly changing current pulse or an RF field.

Eliminating these loops is not realistic for the most part. Minimizing their area, on the other hand, is a good strategy. First, use the minimum cable length. For example, if you have two pieces of equipment six inches apart that are connected with a six-foot cable, most of which is stuffed be-

Figure 5.8 — Many inexpensive audio and computer accessories do not have metal enclosures and cannot be effectively connected to the RF ground plane or bonding bus. For this equipment, place the equipment and any attached cables directly on the ground plane.

hind the operating desk, replace that cable with a shorter one and coil up any extra length as discussed in the previous section.

Where several cables run in the same general area, use wire-ties or a cable tray to hold them close together. Single loops made of stiff wire are an inexpensive way to keep cables together. Run the cables along the RF bonding bus as in Figure 5.9B, further reducing loop area. If you have installed an RF ground plane, run the cables directly along the surface as practical.

Another technique is to use twisted-pair cables to reject low-frequency fields. As described earlier, the balance provided by good-quality twisted-pair cable rejects these signals over a wide frequency range.

Figure 5.9 — Each conductive path through enclosures and cables creates a loop (A) that can pick up magnetic fields. Use short cables and minimize loop area (B) by holding them together with cable ties or in a cable tray and close to the RF bonding bus or ground plane.

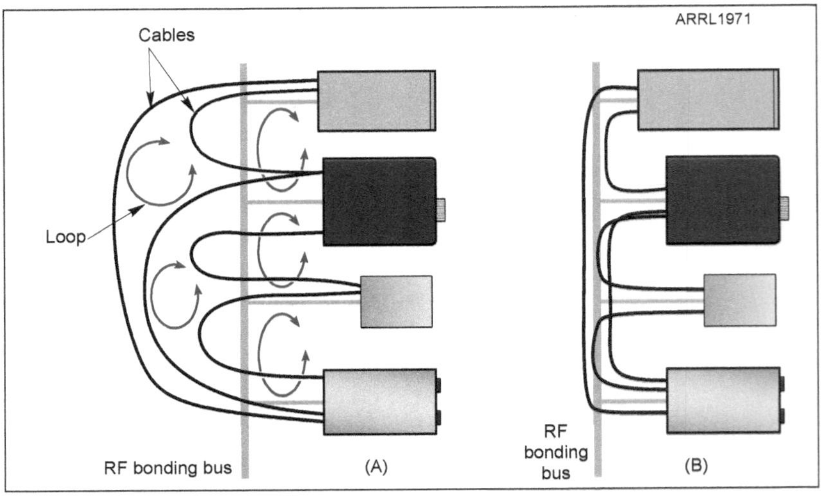

5.4 Blocking RF Current with Chokes

Equalizing RF voltage around your station is hard to accomplish completely. As a result, RF current will flow whenever the voltage "over here" is different than the voltage "over there." You can isolate your feed lines from antennas by using chokes, but the station wiring will still act like an antenna and pick up the radiated signal. After taking reasonable steps to minimize voltage differences, RF current may still be present and causing problems. (See the "Proper Shield Connections" section at the beginning of this chapter.) The next step is to block the current — which is why this section is included in a book on grounding and bonding!

Just as for dc current, the way to block RF current is to put high impedance in its path. That can dissipate some of the current's energy as heat, reduce the current's amplitude, or both. The impedance can be from inductance (inductive reactance or X_L, see the section below on Inductive Chokes), by a tuned circuit that has a very high impedance at its resonant frequency, or by resistance. All of these methods are called a *choke* or *RF choke* since they "choke off" current flow.

• Inductance is created simply by coiling up a coaxial cable. Signals inside the cable are unaffected.

• A cable passed through a ferrite core creates impedance on its outer surface that can be primarily inductive or resistive, or both, depending on the type of ferrite material. Signals inside the cable are unaffected.

• "Sleeve" baluns ¼ wavelength long create a high impedance at one point on the outer surface of a feed line or support mast. These are only intended to work on one band, however!

Of the three, by far the most useful to the station builder trying to deal with RF is the ferrite core. Rolling up a cable to create some inductance can be useful for portable or temporary installations or to troubleshoot RF-related problems. Sleeve baluns are typically used at the antenna and so will not be discussed here. (See *The ARRL Handbook* and the *ARRL Antenna Book* for examples of sleeve baluns as well as other types.)

As a general guideline, for a blocking technique to be effective it should present at least 1000 to 5000 Ω. (The oft-referenced 10 times feed point impedance value is a guideline to avoid antenna pattern distortion.) Transmitting applications or very strong RF fields may require more than 5000 Ω. In a particular situation, you may find that more impedance is needed but most of the time, impedance in this range will suffice to reduce RF current substantially.

Another concern is whether the impedance should be inductive or resistive. A resistive impedance is almost always preferable because it is less affected by frequency and does not add to or cancel with reactance. If the impedance you create is inductive and the impedance of the conductor carrying common-mode current at the location of the inductance is capacitive, the two can partially cancel because they act as if they were connected in series. A resistive impedance makes the choke's effectiveness less affected by feed line length.

Inductive Chokes

Inductance can be created by winding up a cable with or without a form. The resulting reactance is $X_L = 2\pi \times f \times L$. However, the stray capacitance between turns acts in parallel with the inductance to create a *self-resonance*. Up to the *self-resonant frequency* (*SRF*), the coil will act as an inductor, and reactance increases along with frequency. At the SRF, the impedance of the coil can be as high as several thousand ohms. Above the SRF, though, the impedance will steadily decrease. The result is a fairly narrow ("high-Q") resonance.

Without knowing the type of cable, the coil diameter, number of turns, and method of winding, it is very difficult to tell the characteristics of the coil. Two coils that look almost the same might have very different characteristics! (See the RF Techniques chapter of *The ARRL Handbook* for more about self-resonance and other characteristics of components at RF.) Being inductive, the reactance of the choke can be cancelled to some degree if the common-mode impedance of the cable at the choke's location happens to be capacitive.

Figure 5.10 shows a typical coiled-coax choke, and a table of designs is given in the Appendix. If the choke is wound consistently according to the table, it will work as intended on one or two bands but is less effective above and below those frequencies. If the choke is intended for use over wider frequency ranges, the ferrite-core designs work better. You may find coiled-coax chokes useful as a temporary inductive choke for a portable station or at ARRL Field Day. Some experimentation may be required to get the best results.

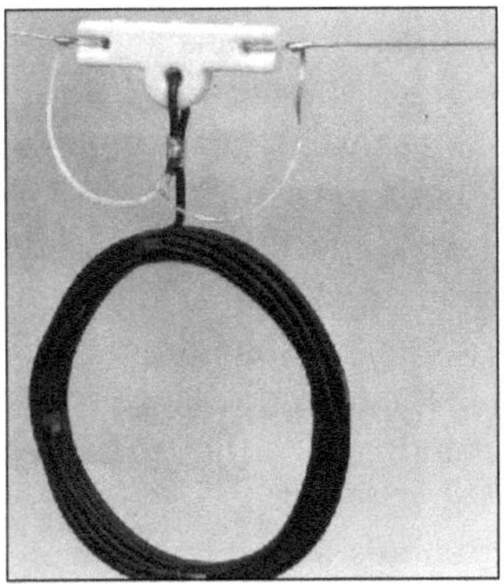

Figure 5.10 — An RF choke formed by coiling the feed line at the point of connection to the antenna.

Ferrite Chokes

A detailed discussion of ferrite is far more than this book can provide. If you are interested in the details of ferrite components, there is more material in the RF Techniques chapter of *The ARRL Handbook* and in K9YC's tutorials (see the sidebar, "In-Depth Tutorials and References" earlier in this chapter). The Fair-Rite Products catalog (**www.fair-rite.com/files1/Fair-Rite_Catalog_17th_Edition.pdf**) also includes a great deal of information about ferrite characteristics. Don't confuse ferrite cores with powdered-iron cores, which are intended for use in RF transformers. Using a powdered-iron core will create some inductance but

Warning: Unmarked Ferrite Cores

Don't use a core of unknown material to make a common-mode choke if you don't know what type of material it is made of. Such cores are unlikely to be effective in the frequency range you are working with. This may lead to the erroneous conclusion that a common-mode choke doesn't work when a core with the correct material would have done the job. There is no consistent industry-wide color coding for ferrite cores. If the seller can't guarantee what type of material is used for a core, don't buy it! When you do buy cores, label them with tape or an indelible marker for future reference since all ferrite looks pretty much the same!

Buying Ferrite Components

Purchasing ferrite cores and beads n small quantities can be expensive. If possible, buy them in quantity from a distributor. This is a good opportunity for club members to share one shipment and get a price break. The appendix of K9YC's tutorial on ferrite (**k9yc.com/RFI-Ham.pdf**) contains recommended guidelines for useful components in a group purchase.

A large shipment — try to purchase by the case — will also be packed by the original manufacturer. Ferrite is brittle and chips easily. A broken or cracked core is not re-usable, even if glued back together. If you receive ferrites that have been shipped loose and not padded, return them as they are likely to have cracks or other damage.

generally creates much less inductance per turn than ferrite.

The impedance of ferrite is composed of both resistance and inductive reactance, both of which change with frequency. Different types of ferrite material (called *mixes*) are designed for use in different frequency ranges, optimized for either "inductive" (inductors and transformers) or "EMI suppression" (resistive) applications. See *The ARRL Handbook* and the K9YC ferrite tutorial for a more complete discussion of ferrite impedance. Ferrite data sheets typically list the best uses for the materials.

Figure 5.11A helps visualize the behavior of different mixes. It shows how impedance varies with frequency of one turn of wire through an identically sized bead (Fair-Rite size 301, about ¼ inch long) made with five different mixes. Remember that the ratio of resistance and reactance varies with frequency as well. Figure 5.11B shows which mixes are recommended by Fair-Rite for EMI suppression in various frequency ranges.

For hams, beads of type 73 material are effective as MF and HF chokes, but they are not available in sizes that fit most amateur coax. Toroid cores of type 75 material, which covers a range similar range to type 73, work well from the 630 meter band through MF and lower HF. Type 31 material is effective as a choke material from upper MF through low VHF. The core material for beads molded into audio and computer cables is typically type 43 for EMI reduction above 30 MHz. Type 61 material can be used for EMI suppression above 200 MHz.

The most useful styles of cores for managing RF in your station are cylinders, clamp-on cores, and toroids. **Figure 5.12** shows some typical examples of ferrite cores. Cylinders are large beads with one hole that are used for single turns of cables and large wires. (Each pass of a wire or cable through a ferrite core counts as one "turn.")

Toroids are a good choice for creating a choke from coaxial cable as

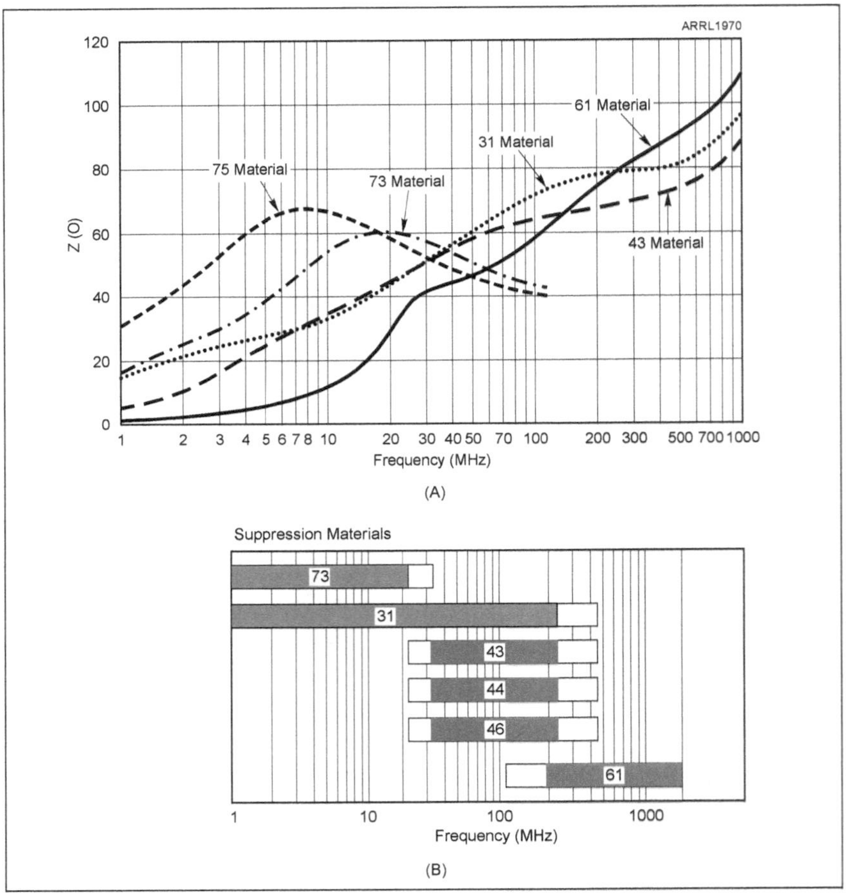

Figure 5.11 — Impedance vs. frequency plots (A) for a single wire passed through one ferrite bead, 3.50 mm diameter × 6.00 mm long (Fair-Rite size 301). Not all mixes are available in all types of cores and beads. The chart at (B) shows frequency ranges for various ferrite mixes used for EMI suppression. Type 73 and 75 material cover approximately the same range. For broadband amateur chokes, type 43, 44, and 46 material are best above HF. [Charts provided courtesy of the Fair-Rite Products Corporation]

Figure 5.12 —Several styles of ferrite cores for common-mode chokes.

shown in **Figure 5.13**. Large toroids can be wound with coax having connectors already installed or with large diameter cable. The most popular toroid size for creating chokes is 2.4 inches OD. Stacks of toroids increase the impedance for each turn of wire. Since turn inductance and capacitance increase linearly with the number of cores, using multiple-core stacks moves the choke's effective operating range lower in frequency.

Figure 5.13 — A transmitting ferrite choke consisting of multiple turns of RG-400 Teflon-insulated cable wound on a 2.4-inch OD toroid core.

If you wind coax through a ferrite core, don't use foam-dielectric cable because the center conductor will slowly be forced through the foam and short to the shield. Also, do not wind the cable tighter than is specified as its minimum bend radius. An exception is RG-400 cable shown in Figure 5.13 with Teflon (PTFE) insulation and the stranded silver/copper center conductor. For legal limit power levels in matched transmission lines, RG-400 can be wound tighter than its specified minimum bend radius if not flexed or stretched once wound onto the core.

Table 5.1 lists several useful type 31 ferrite cores from Fair-Rite Corporation that are available through electronic distributors. Clamp-on cores are basically large beads that have been sawn in two. They have a plastic shell that snaps

Table 5.1
Useful Ferrite Cores (Type 31)

(Measurements are in inches)

Fair-Rite P/N	*Shape*	*I.D.*	*O.D.*	*Length*
0431164181	Clamp-on	0.5	1.55	1.22
0431173551	Clamp-on	0.74	1.15	1.65
0431177081	Clamp-on	1.0	1.7	2.2
2631803802	Toroid	1.4	2.4	0.5

together to hold the core's flat surfaces together. These are used when connectors have already been installed on the cable or if the cable can't be disconnected. It is important to hold the core surfaces tightly together. If the core surfaces separate even a small amount, the core's impedance is reduced drastically. (See K9YC's ferrite tutorial for part numbers of different cores.)

This book's Appendix includes some tables showing recommended sizes and configurations of coiled-coax and ferrite chokes with different numbers of turns.

Placing Chokes

Determining where a choke should be placed is something you'll learn by experimenting. Every station is a little different. Start by determining which equipment is affected and which connections are carrying the RF current causing the problem. Add the choke and re-evaluate. A step-by-step approach to troubleshooting is almost always better than randomly trying different things or "shotgunning" by trying lots of things at once.

Here are some general guidelines for applying chokes:

• Begin dealing with the problem by bonding equipment enclosures together or to some kind of bonding bus or ground plane as described in the previous section.

• Place chokes as close to the affected equipment as possible, whether you are trying to keep RF from getting in or getting out.

• The most useful clamp-ons are 0.75-inch I.D. for small-to-medium cables and 1-inch I.D. for larger cables.

• Multiple chokes can be added "in series." An HF choke added to a VHF choke blocks current over a wide range. Place the choke designed for the highest frequency closest to the equipment.

• Long cables may need one or more chokes along their length to keep them from picking up RF. Place chokes at both ends of cables longer than a quarter wave at the highest frequency of interest.

• When adding chokes experimentally, leave them in place when adding more to other cables; RFI coupling is often a combination of multiple causes.

• If attaching a clamp-on core to a coaxial feed line causes a change in an SWR measurement, RF current on the shield is upsetting the measurement circuits. Add chokes at the SWR instrument or along the feed line.

• Wind both conductors of an ac or dc power cable on or through the ferrite core. This prevents saturating the core by the magnetic field from the large dc or low frequency current.

• You can't place ferrite chokes on an open-wire feed line or wind it into a coil without affecting the differential-mode signals in the line. If common-mode current on an open-wire or window line is causing problems, the K9YC cookbook designs wound with wire pairs may be helpful. They should be placed at the antenna feed point (preferably) or the station entry point and significantly de-rated for power handling.

• Traditional two-winding transformers with an inductive (low-loss) ferrite core and separated windings act as isolation transformers, primarily for receiving applications. The lack of a direct connection between the primary and secondary windings also blocks common-mode RF current paths. At MF and HF, cores made of type 61 material are good for moderate power levels and for higher power, type 67 material.

Additional material on troubleshooting RFI and controlling RF noise is available in *The ARRL Handbook's* chapter on RF Interference, and in the K9YC tutorials.

5.5 Miscellaneous Topics

Detuning an RF "Hot Spot"

When you are setting up a portable or temporary station and can't do a complete job of bonding, you will occasionally encounter a "hot spot" where RF voltage is high enough to cause an RF burn or some other unwanted effect.

If touching the equipment, such as an antenna tuner or feed line, causes a change in the way it behaves, that location is sensitive to loading and is probably a high impedance point in the station. Whatever the reason, you can use a resonant wire to lower the RF voltage.

Remember that peaks and nulls of RF voltage are ¼-wavelength apart and so are the maximum and minimum impedances along any conductor. We can use this to our advantage with a wire that is ¼-wavelength long. If one end of the wire is not connected to anything (an open circuit) the other end will present a low impedance to whatever it's connected. This happens only on the frequency at which the wire is ¼-wavelength or any odd multiples of ¼-wavelength long (¾ λ, 5⁄4 λ and so on).

Cut a piece of wire ¼-wavelength long at the frequency of interest. Use the formula 246 / f (in MHz) = L (in feet), and reduce it by 2 – 3% if the wire is insulated. The tuning is not as critical as it is for an antenna, so you only have to be close. Attach one end of the wire to an antenna analyzer with the other end not connected, and then look for the lowest frequency at which the wire shows a low-impedance dip. This is the frequency at which the wire is approximately ¼-wavelength long.

Solder or crimp an alligator clip or terminal on one end and thoroughly insulate the other end. Attach the non-insulated end to a convenient spot on the equipment where the hot spot is. (Placement is generally not critical.) Run the wire out of the way and place the open end where it won't be contacted unintentionally since it could present a fairly high RF voltage. This should greatly reduce the RF voltage on the equipment to which the wire is attached. The hot spot may be at the open-circuit end of the wire you just added, or it may have moved somewhere else in the station.

What you have done is changed the electrical configuration of your station by adding another conducting element. The station wiring is still picking up just as much RF as ever, but adding the ¼-wavelength wire has moved the hot spot. The technique is similar to curing an unwanted mechanical vibration by adding a weight at just the right location to set up a canceling vibration.

In some situations, it may become necessary to detune hot spots over a wide range of frequencies. In the HF range, the MFJ-931 and MFJ-934 "artificial RF grounds" are often used. These are similar to antenna tuners that can be adjusted to present a low impedance at the terminals connected to the station equipment. These can be very useful for portable and temporary stations where a comprehensive approach is not possible.

RF Noise Reduction

The common-mode RF current flowing on the outside of coaxial cables is not all from the transmitted signal. The shields make good receive antennas for picking up other local signals and noise. There are many noise sources in the typical residence, and neighborhoods are getting noisier all the time. Those signals will be picked up by feed lines and connecting cables. When they encounter an antenna connection or unshielded equipment, the signals enter the feed line as a differential-mode signal, just like the desired signals.

Once inside the feed line as differential-mode signals, noise and other interference cannot be blocked. The noise signals must be prevented from entering the cable using chokes at the antenna end of the cable. The technique is the same as for blocking RF signals picked up from the transmitted signal. You can read more about reducing noise that is picked up on feed lines in the papers by K9YC referenced earlier and by Chuck Counselman, W1HIS, at **www.yccc.org/Articles/W1HIS/CommonModeChokesW1HIS2006Apr06.pdf**. (W1HIS's paper was written before type 31 material was widely available but the general approach of blocking common-mode noise remains useful.) K9YC's tutorial on RFI listed earlier in the sidebar, "In-Depth Tutorials and References," along with his article "Build Contesting Scores by Killing Receive Noise" (**k9yc.com/KillingReceiveNoise.pdf**) presents an extensive discussion of shields and shield connections. For additional guidance about the process of RF noise reduction, see the article "Tracking RFI with an SDR One Source at a Time" by Alan Higbie, KØAV, in the Mar/Apr 2021 issue of the *National Contest Journal*, which is available to all ARRL members online.

Hams aren't the only ones with RF noise problems. Two extensive studies of RF noise mitigation were also conducted by the US Navy as part of the Signal-to-Noise Enhancement Program (SNEP). The authors of these studies were all radio amateurs. "Power-Line Noise Mitigation Handbook for Naval and Other Receiving Sites" and "The Mitigation of Radio Noise and Interference from On-Site Sources at Radio Receiving Sites" are both available from ARRL. Scroll to the bottom of "**http://www.arrl.org/radio-frequency-interference-rfi**" **www.arrl.org/radio-frequency-interference-rfi** to the section, "Naval Postgraduate School RFI Handbooks."

Random-Wire Antennas

Another common situation is an end-fed "random wire" antenna with a high impedance that is connected directly to the station transmitter or antenna tuner. This type of antenna system is often used with a "counterpoise" (a piece of wire connected to the return terminal of the transmitter or antenna tuner). The antenna itself consists of everything — from the end of the counterpoise to the free end of the antenna, plus all of the equipment, the connections between them, and of course, the operator. This also explains why the results from directly-fed antennas can be inconsistent, because there is so much variation in what the antenna actually is!

Applying RF power to the antenna creates a hot spot at the antenna feed point. In this case, adding a ¼-wavelength counterpoise wire to the transmitter or tuner enclosure will lower the voltage on that equipment. (See "Detuning an RF 'Hot Spot' earlier in this section.) The additional wire is actually part of the antenna. Adding the wire effectively moves the antenna's feed point to a lower-impedance point. If a feed line is connected to the antenna and counterpoise, adding a high-impedance choke to the feed line at the antenna may also help by blocking antenna current from flowing on the feed line.

Confirming It's NOT an RF Problem

Diagnosing RF issues sounds easy — if it's a problem when you transmit, it's an RF problem, right? Not for sure! For example, when voltage becomes too low, transceivers can do some strange things! Before chasing an imagined RF gremlin, make sure you don't have one of these simple problems that can cause complex symptoms:

• Remember that "12 V" radios are usually specified to require a narrow range around 13.8 V, typically ±15%. Lead-acid 12 V batteries are often too low in voltage for proper operation unless they are being charged. New battery chemistries such as LiFePO4 remain in the 12–13 V range throughout their discharge curve.

• Power supplies often sag a bit under load, so be sure the supply's output voltage is acceptable when transmitting at full power. Voltage drop in connectors, dc power connector strips, fuse holders, or cables can also cause erratic operation. Use heavy, short wires to supply dc power to the transceiver whether from a power supply or vehicle battery.

• Sudden changes in load, such as on voice peaks or when keying the transmitter can result in short power supply overshoot or dropout transients. An oscilloscope will show short variations in voltage that can cause equipment to operate improperly.

• Loose connections or fuses, or bad cables can create a big voltage drop that looks like a bad power supply. Measure voltage right at the equipment when transmitting to be sure the cable and connectors are OK.

• Using the vehicle frame or body as a dc power return can result in sudden changes in voltage at the radio when the same path is used by some other electrical system that draws a lot of current. Connections from different parts of the frame or body may not be continuous to the battery, leading to erratic changes in voltage. This connection can also upset vehicle controller operation. (See the Chapter 6 section on mobile stations.)

• Accessories or software might act as if they are being interfered with when you transmit, but they might just be configured incorrectly. For example, if the radio is set for split frequency operation, the transmitting VFO or accessory controls might be on a "wrong" band.

• Transmit into a dummy load with a known good coax jumper. If the problem goes away, the cause is RF from your antenna being received by station wiring, or coupled to station wiring by common-mode current on the antenna feed line.

• Poor shields on coaxial cable can cause a variety of problems. Verify that the cables have high-quality shields. Carefully check every shield connection of every cable at every connector. Coaxial cables are often flexed and the shield connection is subjected to the most stress.

• Poor quality RF connectors and adaptors or improperly installed connectors can create many hard-to-troubleshoot problems. Purchase good quality connectors and use the proper technique to install and weatherproof them.

Chapter 6

Good Practice Guidelines

The preceding chapters describing the basics are very nice, but what can *you* actually do in *your* station? Every station is a little different, so a cookbook-style approach is not practical. It is not uncommon for your constraints to make it difficult to follow much of the advice given in the previous chapters. Limits on money, space, and time, while complying with building codes or rental agreements, require compromise and improvisation.

With that in mind, the goal of this chapter is to give you more tools for your "toolbox" of recommended techniques and practices for building and maintaining your stations. Use these examples as guidelines. Where additional detailed material is available online or in a book, links or titles will be provided. More resources are available on this book's web page, **www.arrl.org/grounding-and-bonding-for-the-amateur**. Additional drawings and explanations are available in any recent edition of an *NEC Handbook*, available from public libraries. The National Fire Protection Association (NFPA) also provides free access to the exact codes online at **www.nfpa.org/codes-and-standards/resources/free-access**. There are numerous resources available in the "Free Stuff" section of the Mike Holt Enterprises electrical training website, as well: **www.mikeholt.com**. The Publications, Technical, and Videos sections are particularly useful to amateurs.

Local Codes and Permits

Regardless of the advice in this or any other resource, your local building codes and standards have priority and may require different materials or techniques. When evaluating the examples in this chapter and planning your installation, take the time to evaluate how you plan on complying with local building codes and regulations. Depending on where you live, the situation can be anything from no permits required at all, to *any* electrical work requiring inspection and permitting. The recommendations in this chapter are not guaranteed to comply with building codes — that is *your* responsibility. Take the time to know the rules or you may make an expensive mistake!

Local codes should be assumed to have been based on sound engineering practices that take into account local circumstances and needs. If you are unsure of whether you are meeting your local code, hire a local electrician to inspect, advise, or install your system. Double-check to be sure you are following any specialized codes for amateur installations, such as grounding and bonding.

You may be concerned about trying to do all of the recommended steps at once. Because of time or expense, it may make more sense to do things one step at a time. A step-by-step plan is suggested in the section on Practical Stations. Taking steps slowly will increase your confidence and the quality of your work.

Before beginning, remember that this chapter is written for small- and medium-sized stations. These recommendations and practices are intended to be within the means and capabilities of most hams. Owners of large stations will find some good ideas here, but by and large they have to solve problems on a larger scale and to a higher level of performance. (Big guns — we would love to hear your ideas and suggestions!)

6.1 Ground Electrodes

This section is about selecting, installing, and making connections to earth or ground electrodes. NEC sections 250.52 and 250.53 cover the various types of electrodes and how they should be installed and connected. You should select the type of ground electrodes and connecting materials based on your station and local requirements. An excellent tutorial on ground electrodes is available online from E & S Grounding Solutions at **esgrounding.com**. Look for "Types of Electrodes" in the "About Grounding" section of the website.

Ground Rods

The most common type of ground electrode is the familiar ground rod. Diameter is not electrically critical (typical sizes are ½ inch and the more common ⅝ inch), but a minimum size and length are likely to be specified by building codes. Almost all rods are 8 feet long (the size specified by the NEC). The effect of length and diameter on ground rod resistance is presented in more detail in Chapter 2 of MIL-HDBK-419A (see the Appendix).

Copper-clad steel is the most widely used ground rod material. Galvanized steel and stainless steel are also available. Check with electricians or local contractors and supply houses to find out which is preferred in your area based on soil type to account for possible corrosion and electrolysis.

The NEC and most local codes are clear — two rods are required, although many older installations with a single rod are "grandfathered." Each rod must have a minimum of 8 feet of contact with the soil.

The rods must also be a minimum of 6 feet apart, and twice the rod's length is a good compromise. Closer spacing reduces the effectiveness of the additional rod for lightning protection. The additional rod not only lowers resistance of the earth connection, but also improves mechanical reliability. The rods must be connected together with a heavy grounding conductor (#6 AWG or heavier).

Since most ground rods are 8 feet long and the NEC requires 8 feet of buried length, you'll have to bury the connection to the rod, as well. Take photos or measurements so you can find the ground rod in the future if you need to access it. A good drawing with notes and part numbers belongs in your station notebook!

Drive the rod completely into the ground to the depth required by your building code, then dig out the soil around the top of the rod so that the connection can be installed and covered back up to protect the connection mechanically. If you are using a clamp, be sure it is listed or rated for direct burial. (As mentioned previously, when discussing ac power or safety, "listed" means rated and certified for that use by a certified agency such as UL.)

The NEC requires that the bonding conductor between the ground rods be buried. The conductor should be buried from 6 to 18 inches deep to enhance the connection with the earth and provide some mechanical protection. If frost heave is a problem in your area, consult a local electrical contractor about what to do for buried ground conductors and whether special permits for the necessary digging are required. If the soil is poor, there are ways to lower ground rod soil contact resistance and improve soil conductivity in the area surrounding the ground rod. One method for doing that is to use Bentonite clay installed in a drilled shaft around the rod. An electrician or electrical contractor will know of companies that can provide the materials and installation service. Do not use salt water or a salt compound on the soil. It will corrode the ground rod and destroy the intended connection.

Installing Ground Rods

What tools are best for installing ground rods? An impact drill with a ground rod attachment makes short work of installing rods in most soil types. Search YouTube for "ground rod driver" and you'll get plenty of examples. However you install the ground rod, it must have tight contact with the soil along its entire length. If it rotates or pulls out easily, it's not making good enough contact.

If you use a hammer or mallet, a 5-pound head is big enough for most jobs. Since the rod is 8 feet long, you'll have to stand on something to start driving it in if you're using a hammer. It sounds trite, but be safe — people injure themselves falling off ladders and stepstools all the time thinking they're only a step or two off the ground, so nothing can happen. Wrong! Having someone steady both the ladder and the rod is an excellent idea. Once the rod is seated a foot or two in the ground, the job gets a lot easier.

Consider using a fence post or T-post driver instead of a hammer.

Once slipped over the rod, the driver never misses, the rod is kept from vibrating, you use both arms equally, and all your energy goes into driving the rod instead of swinging a hammer. When the top of the rod gets close to ground level, you'll have to switch to a hammer.

It is preferred to drive the rod in vertically, but if you encounter rocks or a layer of packed "hardpan" soil, it is acceptable to drive the rod in at an angle. The angle should be no greater than 45°. **Figure 6.1** shows some options. Ground rods can be buried in a trench at least 2½ feet deep for the length of the rod.

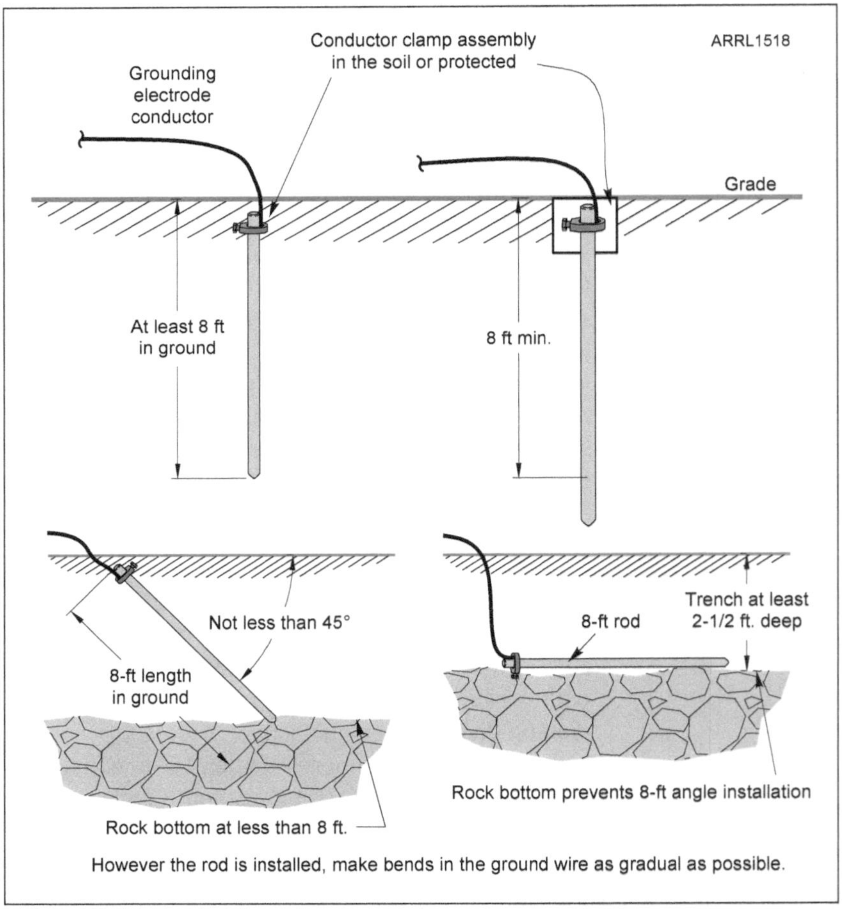

Figure 6.1 — Various approved ground rod installations. The rod must have at least 8 feet of contact with the soil along the length of the rod. However the rod is installed, make bends in the ground wire as gradual as possible. [Based on information in the *NEC Handbook* published by the National Fire Protection Association]

There are many videos online showing different ways of installing ground rods. If you haven't installed one before, it would be worthwhile to watch a few videos to get some ideas of how to (and how not to) do the job.

Ground Conductors

The *minimum* size for ground conductors is #6 AWG wire; see your local code regarding solid or stranded. Some types of ground conductors are required to be heavier, such as connections to Ufer system grounds in the next section which require #4 AWG. Tower ground connections are often as heavy as #2 AWG. Heavy-gauge tinned copper is a good choice for ground rings, connecting radial screens, and for buried ground systems. **Table 6.1** shows recommended ground conductors. Soft copper refrigeration tubing, often available as scrap, can be substituted for heavy wire, then flattened and drilled to form a terminal on the end. If you have questions about the type of material to use, ask a building inspector or electrician about what works well and is approved in your area.

Since ground conductors and connections are often buried, be sure to use terminals, clamps, and fittings that are rated for direct burial. Welded connections discussed below are excellent. Compression fittings that require a multi-ton compression tool are also available.

As discussed elsewhere in the book, heavy, wide strap works well as a ground conductor. The *minimum* size for external lightning protection ground conductors is 1½ to 2-inch wide, 16-gauge (48 oz weight per square foot) copper strap. See the Lightning Protection chapter for a discussion of grounding an SPGP with multiple feed lines attached. (Thinner strap such as 20- or 22-gauge can be used inside the station.)

Flat-braided strap may be used inside and where equipment must be free to move or vibrate. Do not use braided strap where it might be exposed to water or corrosive chemicals. Braid may have to be used in mobile applications as discussed in that section later in this chapter. If so, braid jumpers with weather-resistant jackets are available from electrical and auto parts suppliers.

Table 6.1
Recommended Ground Conductors

Size	*Application*
#14 AWG wire	Operational grounding (see text)
#6 AWG wire	Minimum size for connection to ground electrodes
#4 AWG wire	Minimum size for Ufer ground systems
#2 AWG wire	Tinned copper, used for ground rings and tower grounds
1.5 to 2 in. 16-ga. strap	Minimum size strap for lightning protection grounding

For multi-purpose ac safety, lightning protection, and RF management connections, solid copper strap is preferred. A guide to different types of sheet and strip copper is available at **basiccopper.com/thickness guide.html**. Thicknesses of common sizes are listed in the Appendix. Copper strap is available from some amateur radio distributors and electrical suppliers. Thin flashing is available from home improvement stores or roofing companies.

Aluminum wire can be substituted for copper wire in many cases but should not be buried in direct contact with the soil. Aluminum is very reactive and dissolves quickly in some soil types. Before you use aluminum wire, ask a local electrician or contractor if aluminum is permitted and if it is an appropriate choice for the local environment. Wherever you have aluminum in direct contact with a different metal, use a clamp designed to connect those two metals or use a stainless-steel buffer shim between the two metals to prevent corrosion.

The NEC also mentions "operational" ground connections. These are connections made for electrical equipment to function or for bonding equipment together. Antenna connections to ground terminals or radial screen wires fall into this category. #14 AWG wire is adequate for these purposes.

Grounding Clamps

Use only approved materials for making connections to ground electrodes. Your local hardware store or electrical supply house will have everything you need. If you plan on doing a lot of antenna or tower work, buying parts and pieces in "contractor packs" can save a lot of money and time-wasting trips to the store in the middle of a project. Consider splitting a quantity order or a spool of heavy wire with friends. Having the right stuff on hand makes it easy to use the right stuff!

The clamps you will use most often are those that make connections to ground rods, tower legs, and support masts. The two styles of clamps shown in **Figure 6.2** are the most common.

- GC clamps — two-part bronze or cast zinc clamps with separate screw connections for the ground conductor. The direct burial version has bronze screws.
- RC or "acorn" clamps — one-part bronze clamp that surrounds the rod and the ground conductor, rated for direct burial

For a complete directory of grounding clamp styles, see the Graybar Electric website (**www.graybar.com**) and enter "grounding clamps" in the search window.

Do *not* use hose clamps or any type of clamp or U-bolt not specifically intended for grounding use. They are not intended for long-term electri-

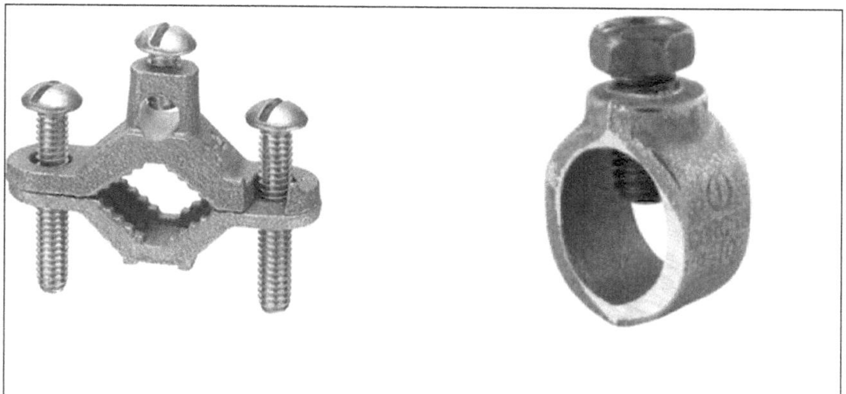

Figure 6.2 — Typical ground clamps. On the left is a type GC clamp used for making connections to pipe or tubing. An "acorn" or type RC clamp designed for ground rods is on the right. For an acorn clamp, the ground wire should be on the side of the ground rod opposite the clamp screw (the bottom of the acorn clamp in this photo).

cal connections because they can loosen or corrode over time. Do *not* use soldered ground connections. Conventional solder connections will fail from the heat generated by lightning current.

If a bronze clamp is being used on a galvanized pipe or tower leg, use a clamp designed for zinc surfaces to avoid dissimilar metal corrosion. A stainless steel shim under a clamp also protects galvanizing as shown in the photo of spark gaps in the Lightning Protection chapter.

Welded Connections

Welded connections are the best for long-term buried connections and for large conductor sizes. Amateurs are probably most familiar with the thermite welding devices known as "one-shots." See the sidebar "Connecting to a Ground Rod" in the Lightning Protection chapter for vendor references. Electrical supply houses also sell the kits. Packages of six or more cost less per kit.

Referred to as "exothermic welding," this technique uses a fast-burning chemical mixture to melt the conductors into a solid, welded connection that can be buried and is very strong mechanically. Typically, a ceramic mold is placed over the conductors, filled with the chemical mixture, and ignited. The process takes just a few seconds! After the weld has cooled, the mold is broken off with a hammer. A typical weld between #2 AWG copper wire and a ½-inch ground rod is shown in **Figure 6.3**. Remember that the rod should be at final depth when the weld is made. Each

Figure 6.3 — A welded ground rod connection between #2 AWG copper wire and a ½-inch ground rod. This connection was then buried as part of a tower grounding system.

one-shot kit for two wires costs around $15 as of early 2021. Newer kits that package igniter powder separately from the thermite are easier to use.

There are quite a number of different molds and configurations for different numbers and sizes of conductors. Use exactly the right product for the connection you need to make. Successful welds depend on conductors being the right size and fitting the mold. Loose fitting molds or improperly seated conductors will not weld properly, and the one-shot kit will have been wasted.

Chris Perri, KF7P, of KF7P Metalwerks (**www.kf7p.com**) has made a video showing all of the parts of a one-shot kit and how to use it. (Search

YouTube for "kf7p exothermic weld.") Carefully follow the instructions and you will be able to successfully make welded connections time after time.

Having an extra kit on hand or some extra welding or ignition powder is handy in case you spill or lose some out of the mold. A sparking tool for ignition is available, and others report success using sparklers or the long fire-starter lighters.

Don't take these devices lightly. The mold glows red-hot in a matter of a second or two — watch the video! Remove flammable material from around the working area and have a fire extinguisher or hose on hand. Wear protective gear, especially glasses and gloves. Do *not* use a cigarette lighter to ignite the mixture, don't smoke around the kits or welding powder, and stay clear of the mold when igniting the mixture.

Ufer System or Concrete-Encased Electrodes

Figure 6.4 and **Figure 6.5** show a typical Ufer ground system formed from the rebar encased in a building's foundation slab. Copper wire can also be embedded in the concrete — #4 AWG is the minimum approved size wire. Rebar must be at least ½-inch diameter, and a direct connection must be made between the embedded rebar and the ground conductor. Ufer grounds have been validated for use as lightning and ac safety grounds, but their performance at higher frequencies is not well-documented. See the PolyPhaser whitepaper "Grounding & Materials" at **www.polyphaser.com/resources/white-papers** for more information about Ufer grounds.

As mentioned in the material on lightning protection, creating a Ufer ground requires welding of all of the rebar and a solid (preferably welded) connection between the embedded rebar and the external ground terminal. If you are unfamiliar with these techniques, the ground system may not perform as you expect, leaving you exposed to lightning damage. Hire a professional to create the necessary connections for you.

It is not realistic to create a Ufer ground "after the fact," meaning on an already-constructed building or tower. You simply don't know how the rebar was installed, and it is unlikely to extend outside the concrete nor it is usually sufficiently bonded. (Ufer grounds usually have to be inspected both before and after pouring concrete, as well.) For new construction, however, the Ufer system may be a good choice if you will have trouble installing ground rods or other buried electrodes. Let your contractor know that you will be building an amateur station so that a connection to the Ufer ground can be created where you expect the station will be.

In the case of towers, it is recommended to install additional ground rods and buried radial wires to supplement the earth connection provided

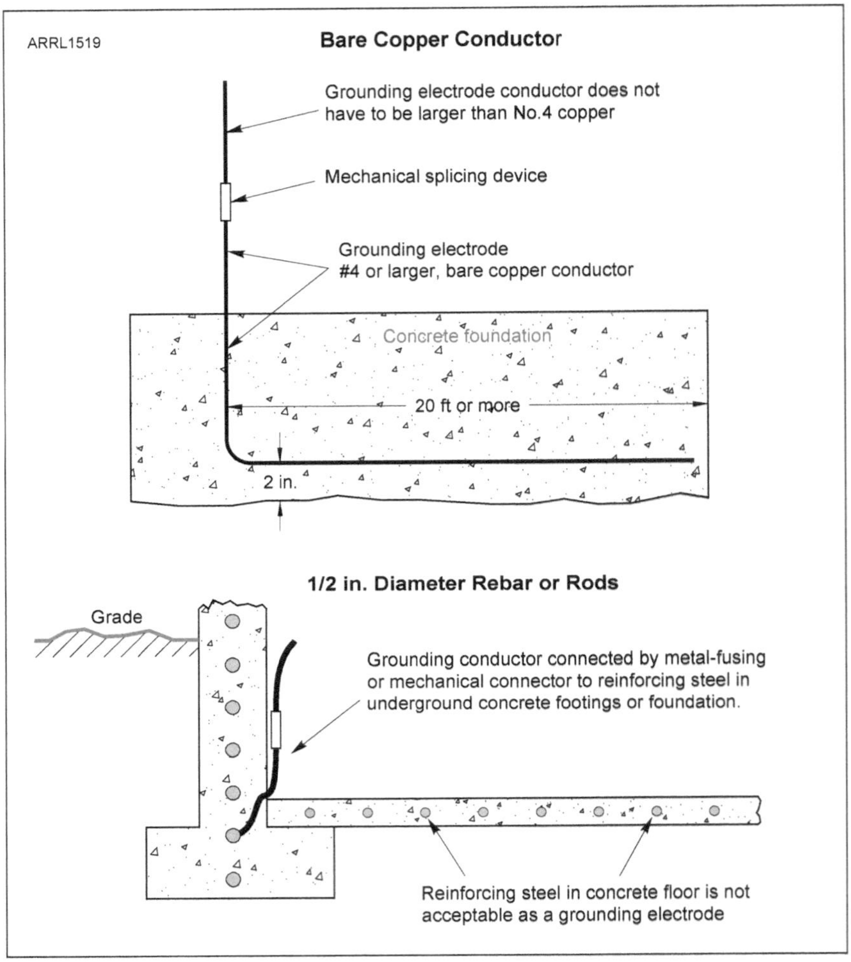

Figure 6.4 — In the Ufer system, the embedded rebar or #4 AWG (minimum) copper must have a continuous length of at least 20 feet and must be part of a footing or foundation. [Based on information in the *NEC Handbook* published by the National Fire Protection Association]

by the tower's base. (See Figures 4.4 and 4.15 in the Lightning Protection chapter.) Adding too much copper to the relatively small tower base can create problems as described in the PolyPhaser paper "Grounding & Materials," referenced above. It is better to provide additional radials to help disperse the charge away from the tower than to rely on electrodes at the base of the tower.

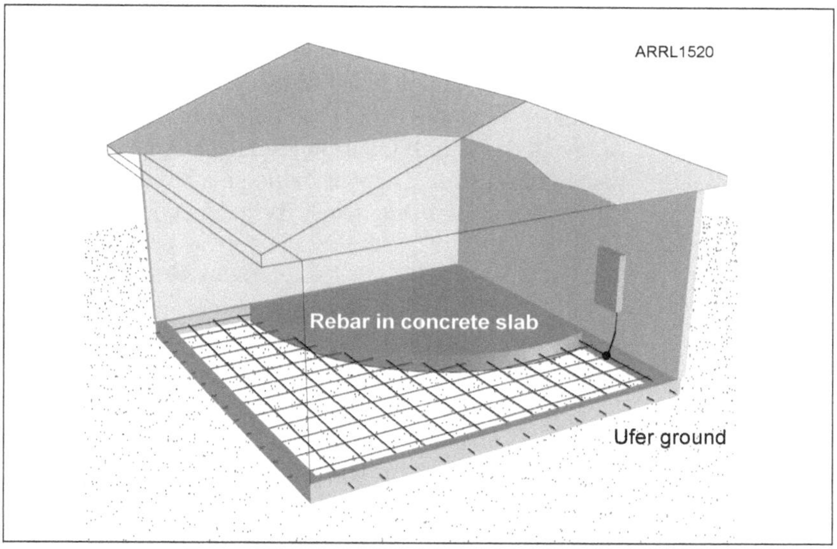

Figure 6.5 — The rebar in a building's foundation slab can be used as a Ufer ground. There must be at least 2 inches of concrete around the rebar, and the grounding conductor must be connected directly to the rebar. The slab must be in direct contact with the soil — no foam insulation or vapor barrier may be used. [Based on E&S Grounding drawing]

Alternate Ground Electrodes

If you do use a pipe connection for grounding, *never* use pipes that contain any flammable liquid or natural gas. All metal piping is required to be bonded together, but only use cold water pipes for ac safety grounding. Also, many water supply systems use plastic pipe with both metallic and plastic sections. Don't assume that a pipe is all metal, particularly a buried pipe. Even metal pipes may be coated or otherwise insulated from the soil. Unless you can somehow verify that the pipe consists of at least 10 feet of continuous metal exposed to the soil, don't use it as a primary ground electrode.

In a residence or other building, plastic and metal components and sections are often mixed. Even a single plastic coupler between two metal sections will make the pipe useless for grounding or bonding purposes. Visually inspect the pipe to be sure it is continuous metal.

For the purposes of lightning protection, you may find that a direct earth connection is not practical for any number of reasons. Use the following alternatives in the order of most to least desirable.

- Building steel
- Stand pipe
- Metal cold water pipe (do not use hot water pipes)

- Metal building skin (if all panels bonded and grounded)
- AC safety ground at an outlet

Even though many of the listed alternatives are unlikely to be effective for bonding at RF, they can help your equipment survive a lightning strike event through SPGP grounding techniques (discussed in detail in the Lightning Protection chapter), which maintain an equal potential between equipment chassis during the strike independently of how the SPGP is grounded.

Tower Grounds

The goal of tower grounding (or grounding of any conductive mast) is to establish multiple short paths to the Earth so that the strike energy is divided and dissipated. You can design the tower base to be a Ufer system as discussed previously or use individual ground rods.

Figure 6.6 is a schematic of a typical tower grounding system. (A tower base Ufer system is not shown.) Connect each tower leg to a separate ground rod. Bond the ground rods together with #6 AWG or larger copper bonding conductor, forming a *ground ring* around the tower base. The ground ring should have a diameter of 6 to 10 feet. The ground ring conductor should be also be buried. (If the ground ring serves as a tie point for an antenna's radial ground screen, burying may not be an option, so be sure it is well bonded to the tower ground radial wires.)

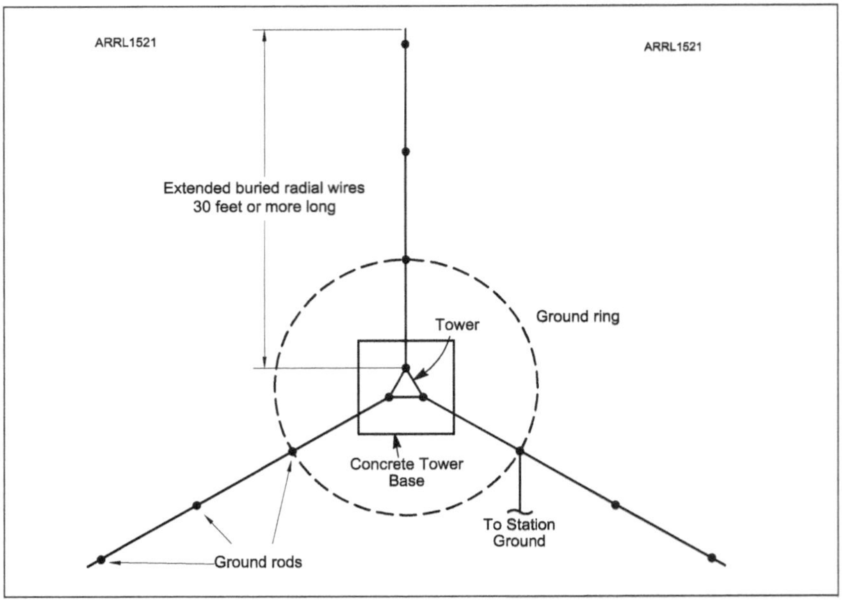

Figure 6.6 — Schematic of a typical tower grounding system (see text). Buried radial wires beyond the bonding ring are optional.

Should I Bond My Tower to My Station Ground?

If the tower is adjacent or close to the building, bond the tower to your perimeter ground using #6 AWG or larger copper wire just as for bonding ground rods together. (See Figure 4.4 for an example.) MIL-HDBK-419A shows towers bonded to building ground systems up to 200 feet away. The bond will work for ac safety grounding and for the lower frequency ranges of the lightning energy spectrum. If the bonding conductor is in direct contact with the soil, it will provide additional ground electrode surface, as well. (See the section, "Antenna Safety Grounding.")

At the guy anchor point, connect each fan of conductive guy wires to a ground rod that is also connected to the tower ground radials. As described previously, do not use solder for these connections. Solder will be destroyed in the heat of a lightning strike.

Because galvanized steel (which has a zinc coating) reacts with copper when moisture is present, use a stainless-steel shim between the galvanized metal and the copper or bronze grounding materials. Several manufacturers make approved grounding clamps for connecting copper wire to galvanized tower legs. Make all connections with clamps and straps approved for grounding applications.

Figure 6.6 also shows buried radial wires extending beyond the bonding ring. The wires should be buried from 6 to 18 inches deep. (These wires are not useful as part of an antenna radial screen, which is typically at or just below the surface of the soil.) These radials act as additional ground electrodes to help dissipate charge in the soil. The buried wire should be #4 AWG at a minimum. Ground rods along the buried radial should be spaced at least 6 feet apart, and twice the rod's length is better. Clamps rated for direct burial or welded connections must be used. These extra radials are a commercial practice included as a guide for amateur stations.

You might have a radial ground screen for a vertical antenna or for a tower or mast being used as a vertical antenna. Those antenna radials can be part of an effective ground electrode, too, along with one or more ground rods. Although they may be on or slightly below the surface, the antenna radials help dissipate charge from lightning strikes in the soil. While there is no standard for using a radial ground screen as lightning protection, a dozen radials or more of bare #14 AWG are reported to be effective and can handle the current without damage. Antenna radials add capacitive coupling to the Earth, which lowers the impedance for the lightning's higher-frequency components.

On rocky ground or even solid rock, a ground system is still possible as described in the PolyPhaser whitepaper "Grounding & Materials" that was referenced in the section on Ufer grounds. The paper recommends the use of solid, flat strap radials so that the strap's sharp edges concentrate the electric field enough to arc onto the surface of the rock or soil, spreading the charge out onto the surface. A Ufer ground is recommended at the tower and guy anchors. The strap can be covered with conductive mortar or concrete.

Antenna Safety Grounding

If your antenna(s) are already grounded by being connected to a grounded tower or mast, there is no need for an additional ground connection. (However, a ground kit will keep any coax connectors at the tower from being welded together.) Bonding around a rotator with a flexible conductor improves the ground connection and may also reduce intermodulation created by the junction between fixed and rotating parts. If the antenna is mounted on a building or non-conductive support, NEC Article 810.21 requires the antenna to be connected to a ground electrode as in **Figure 6.7.**

The wire used to bond the antenna to ground must be at least #10 AWG copper (#8 AWG aluminum is also acceptable). Note that the antenna grounding conductor rules are different from those for the regular electrical safety bonding, or for lightning dissipation grounds. The wire can be stranded or solid.

Run the wire in as straight a line as practical, fasten it in place secure-

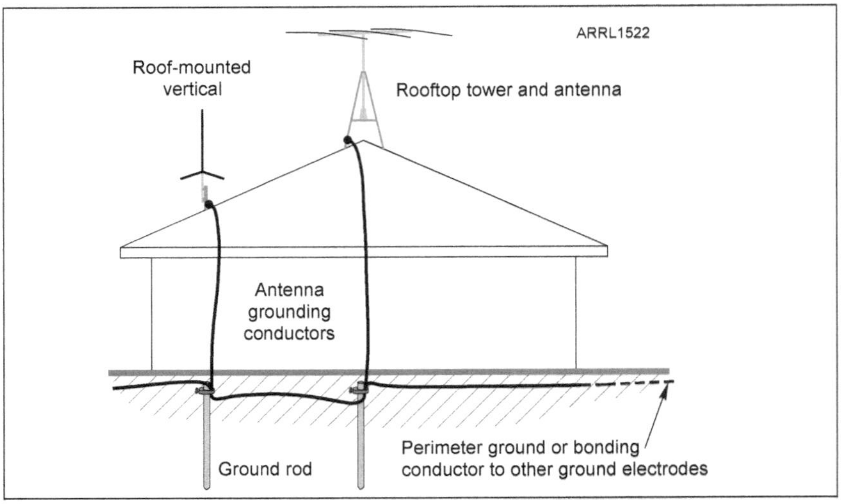

Figure 6.7 — An antenna not connected directly to a ground tower or mast must be grounded separately. The figure shows a typical rooftop antenna tower and a roof-mounted vertical. Each is connected to a ground rod, and the ground rods are bonded to all other ground electrodes by a perimeter ground.

ly (remember the mechanical forces present during a lightning strike!), and protect it from being damaged accidentally. The wire can be insulated or bare. The NEC allows the wire to be run inside or outside the building as necessary, but outside is clearly a safer choice. For receiving antennas, RG-6 coax is available with an integral grounding wire molded into the jacket. This wire is for electrical power and safety grounding, *not* the lightning protection ground. All external antennas must have a lightning protection ground as described in the previous paragraph. For more information, browse to **www.mikeholt.com**, click the search window, select "Website," and enter "satellite dish grounding."

6.2 AC Power

AC wiring practices are well covered by how-to guides available from your local home improvement store and online. A suitable guide is provided in the AC Power System Grounding and Bonding chapter's sidebar, "A Recommended Home Wiring Reference." The slideshow "Power Grounding, Bonding, and Audio for Ham Radio" by Jim Brown, K9YC, offers a detailed presentation on ac power and grounding for hams, including audio circuits. It is available at **k9yc.com/GroundingAndAudio.pdf**.

The notes in this section will help you with common issues in building or upgrading an amateur station. They make a few assumptions about the ac power wiring to your station:

- Correctly wired branch circuit(s) and service entry panel
- Enough capacity to supply your station safely
- Properly installed ac safety ground electrode and the single neutral-to-ground bonding jumper as specified by the NEC

Basic Checkout

Let's start with the obvious. Every piece of ac-powered equipment that is not designed for ungrounded operation *must* have a three-wire, grounded power cord. No broken-off ground pins on power plugs or snipped-off green wires are allowed! Replace or repair defective power cords or plugs.

Switch to Safety

If it's within your budget and available space, install a lockable disconnect switch or circuit breaker in its own enclosure that supplies all ac power to the station equipment. Another possibility is a small sub-panel with a few circuit breakers. Either is a better and safer station master switch than relying on the light-duty switches in garden-variety power strips. And it's a lot easier for anyone else in your home to turn off power should there be an emergency of some sort.

Evaluate Circuit Loading

Assuming all of the outlets feeding your station are on the same branch circuit — a common situation — the next step is to find out if it can safely handle the load. The NEC (Article 220) limits the maximum sustained (continuous) current load on a circuit to no more than 80% of the breaker's rating. For example, if the circuit is protected by a 15 A breaker, your equipment should draw no more than 15 × 0.8 = 12 A on a steady basis. A 20 A breaker can handle 20 × 0.8 = 16 A on a sustained basis. If the load is approaching the maximum for a circuit (don't forget lighting or non-radio appliances) you should install at least one other branch circuit. Most stations are well within the capacity of a single 20 A branch circuit. Installing a 20 A circuit costs little more than one rated for 15 A. If you do add a 20 A circuit, using #10 AWG conductors lowers voltage drop from conductor resistance, especially if capacitor-input power supplies that draw current at voltage peaks are used.

If you have an amplifier capable of putting out 1 kW or more, it should be powered from its own circuit breaker, since powering it from a 120 V circuit leaves little for any other equipment! Most tube amplifiers are capable of being powered by either 120 V or 240 V. Newer solid-state amplifiers with 48 – 52 V dc switching supplies can also operate on 240 V. Powering these loads from 240 V instead of 120 V reduces current draw by half. The power supply output will drop or "sag" less under load at the higher voltage, as well. A dedicated 240 V circuit for amplifiers and other high-power loads is recommended.

Check the resistance between the power cord ground pin and the metal enclosure of the equipment — the ground lug inside might not be making a good connection due to a worn cord, loose connection, or even paint on the chassis.

An ac hot or neutral connection must *never* be connected to any exposed metal. Vintage equipment with a two-wire power cord should be treated with care — see *The ARRL Handbook*'s chapter on Troubleshooting and Maintenance for guidelines on working with older equipment. Beware of older amplifiers and appliances powered by 240 V that obtain 120 V from a connection between hot and ac safety ground.

Finally, inspect the outlets that supply your station. Verify that all wires are connected properly at the outlet. Replace any worn outlets. Identify and label the circuit breaker(s) supplying power to the station for each outlet. Note the circuit breaker capacity.

A comprehensive figure showing the proper connections for a wide range of 120 V and 240 V ac plugs is provided in the Appendix.

Power Distribution

In a fully lightning-protected station, ac power must be supplied by a protected circuit from the SPGP. Sharing a single 15-A branch circuit is quite possible for a transceiver, an RF amplifier up to 500 W output powered by 120 V, and station accessories such as a modern computer and LCD monitor. (See **Figure 6.8**.) An RF power amplifier powered by a 240 V circuit should have the ac safety ground within its outlet bonded to the safety ground wires used by other station ac outlets.

Figure 6.8 — Sharing the branch circuit ground conductor reduces small voltage differences from leakage current in the equipment that causes hum or buzz in low-level audio circuits. Bonding equipment together for RF management or lightning protection also minimizes these low-level differences. The perimeter ground bonding the SPGP and the ac service entry ground electrode further reduces the shared voltage drop. Using a protected outlet mounted on the SPGP is preferable to using the branch circuit outlets directly.

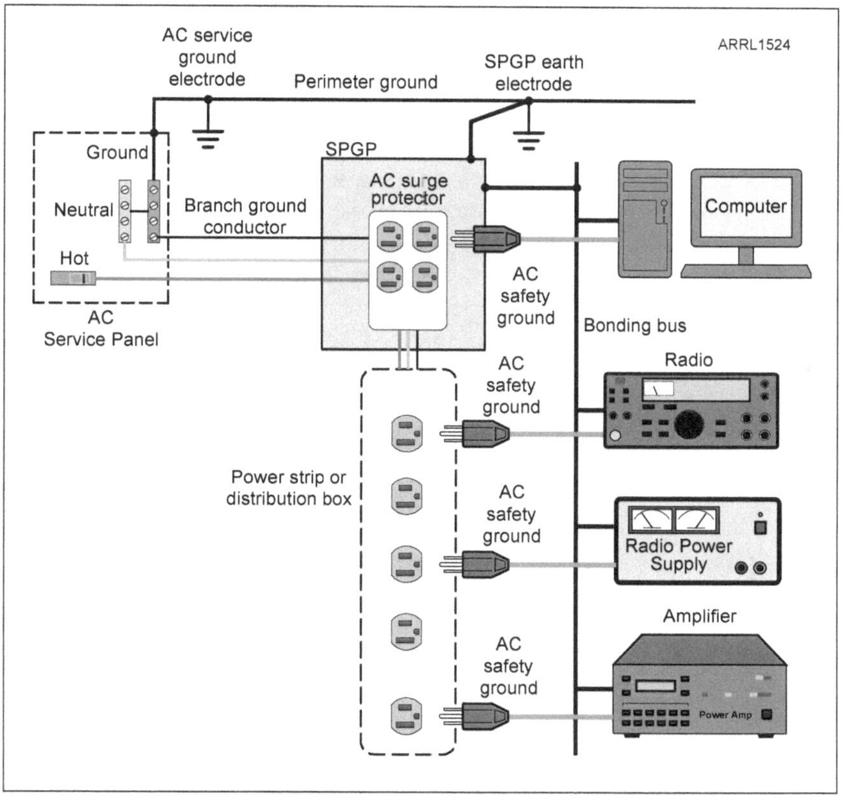

GFCI in a Box

There are situations in which we don't always have as much control over safety as we'd like. The GFCI-in-a-box outlet helps restore safe operation. It consists of a GFCI-protected (Ground Fault Circuit Interrupter) duplex outlet in a metal junction box with a heavy power cord that plugs into any regular outlet (see photo). Models are available with weather-resistant covers or plugs. This is great for workbench use when testing equipment as well as working outside. You can put it in your Field Day kit and use it for portable and temporary stations, too.

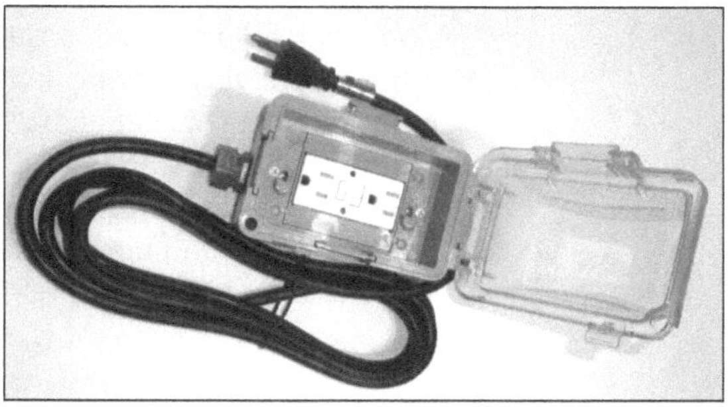

A protected circuit will need to supply multiple devices. Most stations have enough gadgets and radios and power supplies that a single outlet, or even a single power strip of a half-dozen outlets, won't suffice. Resist the urge to use outlet splitters and cheap power strips!

Figure 6.9 shows a typical power distribution system that can be supplied from a single protected outlet. You can build your own heavy-duty, high-quality power distribution box as in **Figure 6.10**. Use metal "back boxes" with high-quality outlets and heavy #12 or #14 AWG wiring. Multiple boxes can be attached together mechanically to support as many outlets as needed. If more than one multi-outlet box is used on the same circuit, use rigid EMT or flexible metal BX conduit between them with all of the metalwork solidly assembled and connected to the ground conductor. A heavy switch can be included to turn the entire package on and off.

Plugging smaller strips into the power distribution box or strip accomplishes two things. First, it allows you to power similar equipment from one strip and be able to turn it on and off independently. For example, a computer, printer, monitor, and speakers could be on one power strip, with your transceiver power supply and rotator control box on another. All equipment must still be bonded together for lighting protection and RF management.

Wherever commercial power strips are used in your station, use industrial-quality products with a metal enclosure, heavy wiring, good quality outlets and a built-in switch and circuit breaker (typically rated at 15 A). Avoid inexpensive plastic power strips. As discussed in the AC Power System Grounding and Bonding chapter, power strips should meet the UL 1363 standard. (This will be displayed on a tag or molded into the body of the power strip.) If the power strip has a surge protector, it should meet UL 1449 (3rd Edition), as well. Some strips and most surge protectors have lighted indicators to let you know the protective components are still functioning.

In some situations, it may not be practical to power all of the station equipment from a single protected outlet on one SPGP. For example, running extension cords to different operating positions. In such cases, install a separate SPGP at each position with its own ac safety and ground connection, preferably sharing a common ac safety ground conductor. There must be no direct connection between the positions, such as accessory power supplies ("wall warts"), audio, or control cables. Any other configuration is likely to result in voltage differences during a surge that will equalize through the equipment. Reconfiguring or relocating equipment is a better strategy.

GFCI Protection

Using ground fault circuit interrupter (GFCI) circuit breakers or outlets to supply power is not required by the NEC. However, given the many op-

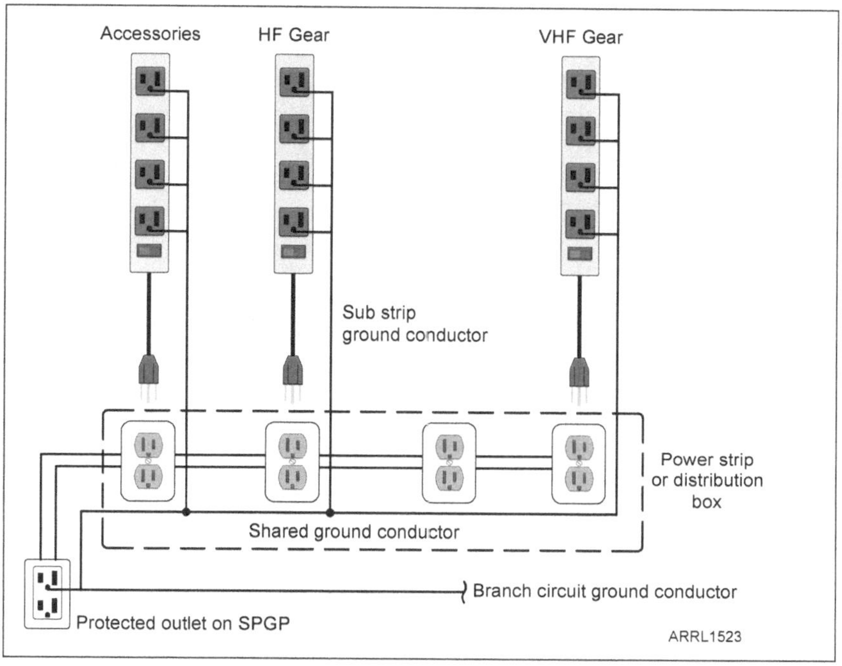

Figure 6.9 — Multiple outlets are needed to supply ac power. Either make your own power distribution center or use a high-quality, heavy duty power strip. Don't daisy-chain the smaller power strips. In this way, all ac-powered equipment shares the same ground conductor on the branch circuit. Computer equipment should be powered from the same outlet(s) as the radio equipment it connects to.

Figure 6.10 — A power distribution center can be homemade by using metal outlet boxes and installing your own outlets, as this multi-outlet box shows. See the text for more details. [Photo courtesy of Jim Brown, K9YC]

portunities for stray current in an amateur station and antenna system, GFCI protection is not a bad idea. You can install GFCI breakers in the ac service panel or install GFCI outlets at the station. If you build your own distribution box, one GFCI-protected outlet per box can also be included. If it is desired to have more than one GFCI-protected outlet in a box, most GFCI outlets have provision for protected feed-through lines for more outlets. Wire the extra outlets to the feed-through terminals of the GFCI device.

When GFCI breakers were a new product, there were some reports of RFI causing them to trip. These problems have been corrected in current products. See the ARRL report on GFCI circuit breakers and RFI at **www.arrl.org/gfci-and-afci-devices**. If you have questions or a report of RFI to a GFCI device, contact the ARRL Lab's RFI specialist. Note that vintage radios often have high enough leakage current to the chassis that a few of them on the same circuit could cause a GFCI to trip.

Bonding for Audio Quality

As introduced in the RF Management chapter's discussion on bonding, hum and buzz can be present and affect low-level audio signals. This can cause problems in microphone audio or low-level audio circuits used for digital modulation. The best way to mitigate these problems is to follow all the rules for bonding, both in the station and throughout the premises as a whole. Especially bond every piece of equipment that's interconnected, chassis to chassis, following the shortest practical path.

AC power for station equipment, including accessories such as audio, computers, controllers, and other devices, should be from the same protected outlet on the SPGP. If there is no SPGP protector, ac power should be supplied from outlets on the same branch circuit that share the same ac safety ground wire back to the ac service panel or from outlets in close proximity whose green wires are bonded together at the outlet.

Using a common safety ground connection and bonding chassis-to-chassis minimizes the leakage current carried on the typically light cable shields by shunting it with the heavy bonding conductors. While that wouldn't be adequate for high fidelity audio, it's fine for ham radio. Note that the requirement for bonding does not include equipment that is located in the room with the station but is not part of the station, such as lighting, or a computer, monitor, or printer not connected to the station.

It is also important to use good-quality audio connectors such as those from Switchcraft and Neutrik. For single-circuit cables, RG-174 miniature coax is a good choice because of its braided shield and because it fits easily in most small connectors. For stereo and other multi-circuit cables, use shielded, twisted-pair cable such as Gepco XB401 or Belden 1901A.

The leakage current is typically "buzz" from the power system triplen harmonics discussed in previous chapters, and from current through line filter and bypass capacitors. The voltage drops from each piece of equipment sharing the common ground conductor add together. The resulting voltage is present on all equipment connected by the common ground conductor. In addition, not every piece of equipment contributes the same amount of leakage current. The resulting voltage differences between equipment enclosures may be only a few millivolts but that can be enough hum or buzz to affect a sensitive audio input.

The recommended solution is to create heavy bonding connections between pieces of equipment, either through a bonding bus (good) or directly from enclosure to enclosure (best). In **Figure 6.9**, the heavy lightning protection bonding conductors do double duty by shorting out any ac hum and buzz voltages between pieces of equipment. In addition, the required bonding conductor between the ac service entry and lightning protection ground electrodes (labeled "perimeter ground") provides a better connection than the branch circuit ground conductor.

Rack-Mounted Equipment

If the equipment is designed for rack mounting, install the equipment on steel mounting rails, cleaning off any paint from the rear surface of the equipment mounting ears and the steel mounting rails. When equipment lacks proper mounting ears, bond from the shielding enclosure to the steel rack rails. Then bond those rails to the operating desk and bond the rack cabinet to the SPGP.

Most amateur equipment, however, is not designed to be rack-mounted. As a result, a bonding jumper (strap or wire) is installed from the equipment enclosure to the rack. A *rack grounding bus* or *RGB* is usually mounted parallel to the equipment mounting rails as shown in **Figure 6.11**. Each piece of equipment within the rack is connected to the RGB with a #14 AWG or larger wire or strap. The RGB itself should be con-

An AC Ammeter Accessory

A clamp-on ammeter that can easily be placed in-line with an ac power lead is a very useful tool. These are available as standalone instruments (the Fluke 323 is typical) or as an accessory for multimeters, such as the Fluke i400. To allow the clamp-on to measure current on any of the ac power conductors individually, modify a short 3-wire extension cord by removing the outer jacket for approximately 6 inches. A meter with low-current ranges is most useful in the ham station — a maximum range of 20 to 30 A max is sufficient. A comparable oscilloscope current probe can show ac leakage currents, as well.

nected to the rack enclosure, usually done with the hardware used to mount it in the rack. The enclosure of each piece of equipment in the rack is then connected to the RGB with a wire or strap.

The rack itself and the RGB (if used) are then connected to the ac safety ground and the station lightning protection ground system. Power outlets in the rack should be in steel boxes bonded to the rack. If the rack is fixed, solid strap or a heavy wire should be used as the ground conductor. For rolling or moveable racks, flat-weave braid is an acceptable substitute indoors, but only if water or corrosive chemicals are not present. Keep the strap as short as practical.

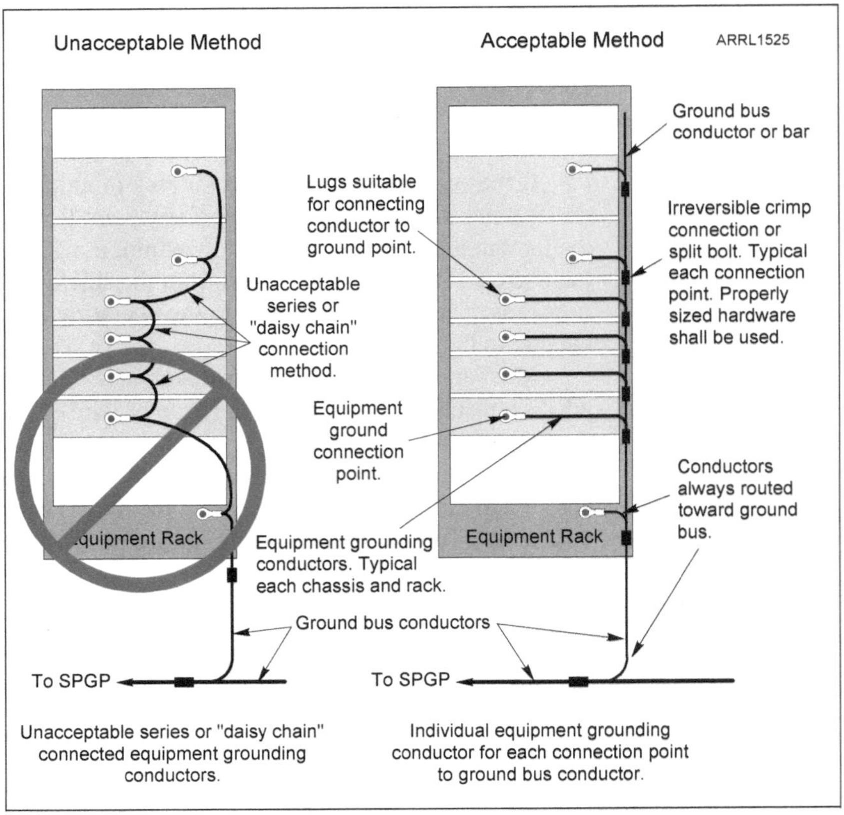

Figure 6.11 — When mounting equipment in a rack cabinet, install a rack grounding bus (RGB). Equipment is connected to the RGB and then the RGB itself is connected to the station ground system. [Based on information in *Standards and Guidelines for Communication Sites*, R56, courtesy of Motorola, Inc.]

6.3 Lightning Protection Planning

This is an important part of the book — devising a lightning protection plan. The following material is from the 2002 series of *QST* articles by Ron Block, NR2B (then KB2UYT). The complete set of articles is posted online at **www.arrl.org/lightning-protection**. Once you have a plan, you can implement it using the techniques presented in this book.

Repeating the caution that every station is different, there can be no "cookbook" for lightning protection. You'll have to follow the basic principles thoroughly and carefully, or your protection plan will be flawed. A flawed plan will not protect your equipment as you expect. Proceed step by step, and apply the information in the preceding chapters to your specific station.

Identify What to Protect

The goal of the planning process is to establish a "zone of protection" within a station, as opposed to the whole house or building. Additional zones may be considered separately. Start by identifying what you want to protect. While you would like to protect everything, create a priority list and work the list from high to low priority.

First on the list are probably the more expensive items associated with your station, usually the transmitting and receiving equipment. What follows depends on just how you enjoy the hobby — the antenna tuner, linear amplifier, accessories, or computer. Further down the list might be the antenna, rotator, and feed line. Each person's list and priority ordering will be different. Pause here and mentally construct your priority list, being sure to include all the elements of your radio station. We will then work through the process of developing your protection plan.

The first step of the plan is to construct a complete block diagram of the equipment in your station, starting with the top priority item. (You will make a separate plan for other areas needing protection.) This is usually simple and straightforward. In some installations, it may be necessary to look behind the equipment to determine precisely the connections between each element. The accuracy of the diagram is important in determining the nature and effectiveness of the protection plan.

Assume the list's top priority item is your transceiver (or transmitter and receiver, if separate). If you have multiple transceivers, list them in

> **Protect Yourself!**
>
> Please note: Just because equipment may survive a direct lightning strike does not mean that you can. You must not be in contact with the equipment during a strike because your body presents a conductive path that probably breaches the protected equipment circle to the outside world. The resulting current could hurt both you and your equipment.

Ethernet or Wi-Fi?

The cables that carry Ethernet wiring around houses and ham stations are a common path for lightning damage — even if they never leave the building. In addition, cables between routers, switches, and computer network interface connections are all common sources of RFI. If you can, change your network connections to a wireless technology such as Wi-Fi and eliminate both problems. Likewise, consider wireless phones rather than wired extensions. This is particularly useful advice for upper-floor and outbuilding stations to avoid more long cables that create paths for lightning surges.

order of value. These are the heart of your station, so make them the starting point of protection plan, which will in turn examine and diagram each element of the station.

Assuming your primary item is a transceiver (we'll refer to separate receivers and transceivers as a transceiver for simplicity) represent it in the block diagram as a single rectangle. Label it with the manufacturer's name and model number.

Next, think about the antenna connection to the primary transceiver. If the connection goes directly to the external antenna, simply draw a line from the rectangle to the edge of the paper. However, if the antenna is connected to the equipment via a linear amplifier, antenna tuner, or a multi-position coax switch, add this (these) as separate rectangle(s) interconnected with the primary radio equipment. The feed line going to the antenna should still go to the edge of the paper. Label the feed line's lowest and highest frequency (MHz or band name), the maximum transmit power in watts (rounded up), and the type of connector and gender (UHF/PL-259 male or N-series male, for example).

Add a rectangle to the diagram for each additional transceiver or amplifier in your station. Be sure to show all interconnections and antenna connections for each of these secondary rectangles. If any of the secondary radio equipment has a direct connection to an antenna, show the feed line going to the edge of the page. Be sure to label each rectangle with the manufacturer's name and model number and each feed line with connector type and gender, frequency range, and maximum transmit power. **Figure 6.12** shows a block diagram for a simple station. (This discussion assumes the station is for HF operation. The same planning process works for VHF+, too.)

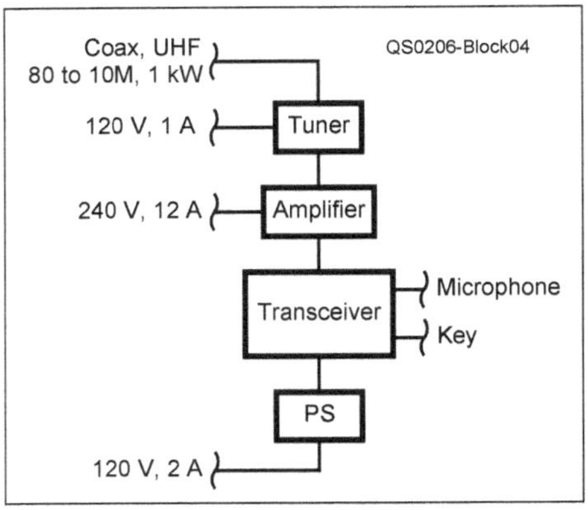

Figure 6.12 — Block diagram of a typical simple HF radio station.

The block diagram should now have a rectangle representing each piece of radio equipment and all accessories in the station. Each of the rectangles should have lines representing the interconnecting cables and feed lines. Each feed line that leaves the station and goes to an antenna or some tower-mounted electronics should be drawn to the edge of the page and labeled.

Take a Detailed Look

Now it is time to examine each piece of equipment represented by the rectangles, one at a time, and to add rectangles to the diagram for any other

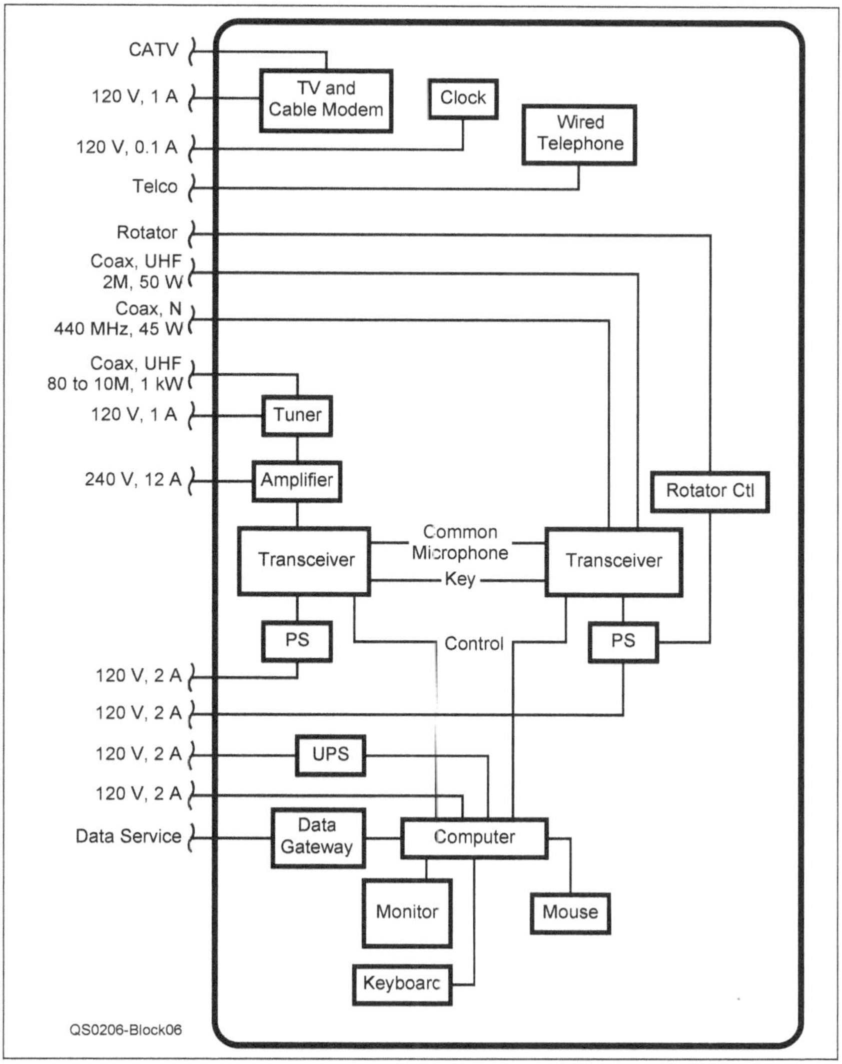

Figure 6.13 — Block diagram of a typical more-complex radio station. The heavy outline identifies the desired zone of protection. Lines that penetrate the outline are the radio station I/O circuits that must be protected.

electronic devices, complete with the electrical connections and interconnections between them. Some of these will be easy and intuitive, while others will require examining the equipment. Every connection must be included — this is important to the integrity of the plan and how you implement it. The only exception is a nonconductive fiber-optic connection.

To complete the diagram in an orderly fashion, pick a rectangle and answer all of the following questions for that rectangle. Then, pick another rectangle and do the same until all of the rectangles have been examined.

1) Is there a connection between this rectangle and any other rectangle? If so, add a line between the respective rectangles and label its function.

2) Is there a connection between this rectangle and a device not yet included on the block diagram? This can include standalone amplifiers, power supplies, computers, audio interfaces, network routers and switches, USB hubs, broadband gateways, and the like. If so, add the new device to the diagram as a rectangle and label it. Then add and label the connections. Repeat this step until all connections from this rectangle to new devices have been completed.

3) Is there an ac power connection required for this rectangle? If so, draw a line to the edge of the page and label it with the voltage and current required. If the rectangle represents equipment with MOVs connected to the ac safety ground, such as a UPS or tower-style PC, the ac power outlet must be directly bonded to the SPGP. Laptops supplied by a 2-conductor power cord are not connected to the ac safety ground.

4) Is there a requirement to supply ac or dc power through a feed line to operate remote switches or electronics? If so, label the feed line at the edge of the page with the peak voltage and current requirements.

5) Are there control lines leaving the rectangle going to remote electronics, relays, or rotators? If so, draw a line to the edge of the page and label it appropriately.

6) Is there ac power leaving the rectangle going to the tower for safety lighting, convenience outlets, crank-up motors, or high-power rotators? If so, draw a line to the edge of the page and label it with the voltage and current required.

7) Is there a connection to a telephone line, data service, satellite TV, or cable TV connection (RF, video, or data) for this rectangle? If so, draw a line to the edge of the page and label it appropriately.

8) Is there a connection to another antenna system such as for GPS, broadcast or cable TV, or a satellite dish for this rectangle? If so, draw a line to the edge of the page and label it appropriately.

9) Is there a connection to other equipment elsewhere in the house or building, such as for network or telephone wiring? If so, draw a line to the edge of the page and label it appropriately. You may want to do a physical

inspection of the room for overlooked phone jacks or other electrical wiring.

Once you have completed the process for each of the rectangles, including all of the new ones that were added, you should have an accurate block diagram of your radio station. It may be prudent to review each rectangle to verify that nothing was left out. Your block diagram should look something like **Figure 6.13**.

Now step back and physically look at the equipment in the station. Has every piece of equipment been reflected in the block diagram? Every metallic item within four feet (in all directions) of the radio equipment must be considered as a part of the radio equipment even if it is not actually connected to it electrically.

If there is such an item that has not been included, we need to carefully evaluate it. Don't forget to include a metal operating desk! An example of such a device could be a simple standalone telephone on the operating desk or a computer system (CPU, monitor, keyboard, and mouse) some part of which is sitting on or near the radio desk.

Nearby devices (telephone and or computer), while not a part of the radio station electrically, are within a spark-gap's reach of the radio station equipment. Therefore, they are considered connected to the station just because of their proximity, and they must be added to the block diagram. Follow the same procedure that you used to add equipment to the block diagram.

Now that the diagram is accurate and complete, draw a circle around all of the rectangles, allowing each of the lines that extend to the edge of the page to cross the circle as shown in Figure 6.13. The equipment represented by the rectangles within the circle is to be protected. All of the lines going from the circle to the edge of the page are called I/O (input/output) lines or circuits. Remember that everything outside of the circle is not a part of the zone of protection. Separate protection zones can be created using this technique.

Now that you have identified all of the I/O lines for the station, each must be protected and each of the I/O line protectors must be grounded and mounted in common, such as on an SPGP. Techniques for protecting each of the I/O lines are discussed in the chapter on Lightning Protection.

Assume Protection is All or Nothing

One word of caution regarding accuracy and attention to detail: You should assume that the protection is "all or nothing." If an I/O line is inadvertently missed, then the protection plan is flawed and lightning energy has a way through your station to a different ground connection. The resulting damage can be just as bad as having no protection at all.

Think Like Lightning

Another part of the planning process is deciding where to put your station in your house, how ac power will be supplied, and where your feed lines and other services will enter the house. It is understood that you may have no control over most or all of these choices! What you can control is to do everything you can to guide lightning's energy into the soil via the ground electrodes — without going through your station!

A lightning pulse can come via your antenna system or from a strike on the power lines, CATV system, telephone lines. A strike can also induce surges on all wiring within a building. **Figure 6.14** illustrates how the SPGP and bonded ground electrodes work together to provide a low-impedance path to dissipate the charge in the Earth. If the bonding conductor between the utility ground electrodes and the station ground electrode is not present, lightning energy will travel through *all* of the equipment within the protected zone to find another ground.

If you can keep your various surge protectors close together where the cables are to enter the protected zone, that will help equalize the voltages on all conductors so that current does not flow between them. Anoth-

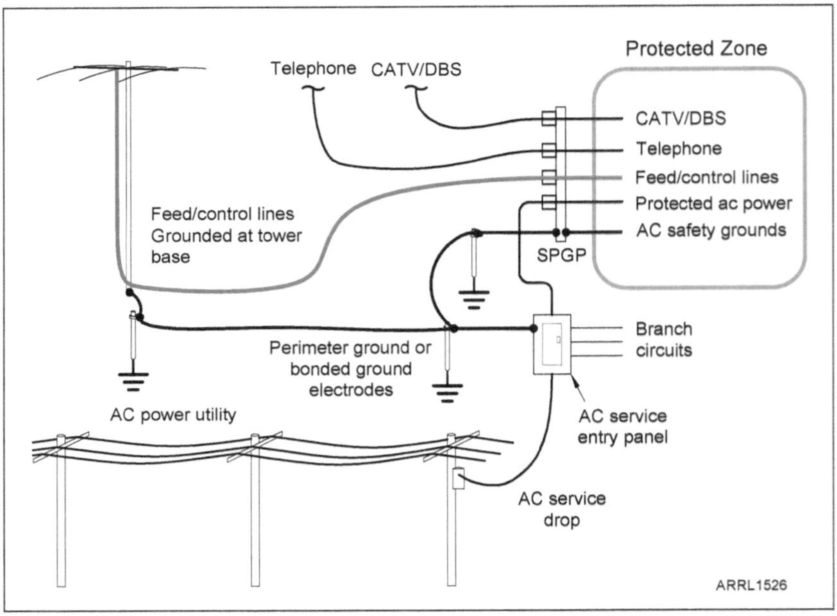

Figure 6.14 — External ground electrodes and heavy bonding conductors between them provide a low-impedance path to the Earth. All I/O lines entering or leaving the protected zone must have some kind of mitigation against lightning. These measures give lightning a path around your station no matter how the energy is routed toward your station.

er reason to keep the protectors close together is so that the protectors act in unison. This prevents surges from arriving at different pieces of equipment at different times, resulting in destructive voltage differences.

On his website, Tom Rauch, W8JI, presents several scenarios for lightning and discusses how he has configured his station. His station may be large, but the many photos and drawings show how he has successfully dealt with lightning. You may get some good ideas from his experiences. The primary pages on his website about grounding are:

- **w8ji.com/ground_systems.htm**
- **w8ji.com/house_ground_layouts.htm**
- **w8ji.com/station_ground.htm**

Inside the station, you can avoid creating paths for lightning by keeping bonding conductors short. For example, if you have more than one branch circuit available to power your equipment, choose the circuit with an outlet closest to the SPGP or ac service entrance and connect the bonding jumper at that outlet. If you are going to run new or additional circuits, locate them so the ground path to the SPGP is short and direct. Based on examples from W8JI's **house_ground_layouts** web page listed above, **Figure 6.15** shows an example of keeping ground bonding paths short by careful placement and selection of ac power and ground electrodes. Mount an ac line surge protector on the SPGP and supply your station from that protected point. Do not distribute ac directly from an unprotected wall outlet or distribution panel; this will invalidate the zone of protection. Any MOV-based surge protector or equipment containing MOVs connected to the ac safety ground conductor must be fed from the protected outlet on the SPGP.

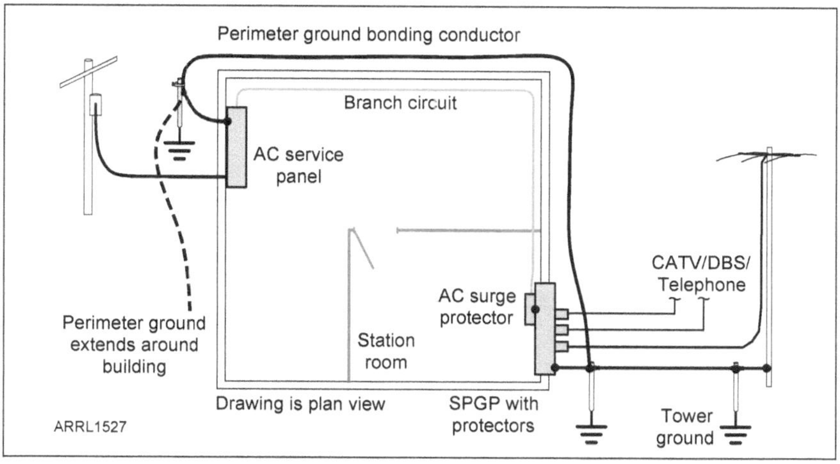

Figure 6.15 — Keep ground bonding connections short. Use outlets close to the SPGP, if possible. The closer together the ground electrodes and protected entry points, the better your lightning protection will perform.

6.4 Managing RF in the Station

An overview of dealing with RF from your transmitted signal was presented in the RF Management chapter of this book. This section will provide a few examples of successful techniques you can use as ideas for your own station.

RF Ground Plane

Figure 5.5 in the chapter on RF Management shows a basic operating table RF ground plane with a bonding bus clamped to it. Aluminum roofing flashing is used for the ground plane and lightweight copper pipe is used for the bonding bus. Copper flashing or brass sheet, copper-clad PC board material, fine-mesh screen — any continuous metal surface will work well. The bonding bus and ground plane must be connected to the ac safety ground either with a direct connection to an ac outlet or via the SPGP.

The ground plane is held to the plastic tabletop with self-tapping sheet metal screws. If you don't want to drill into the tabletop, the flashing can be attached with double-sided tape, glued to thin cardboard or poster

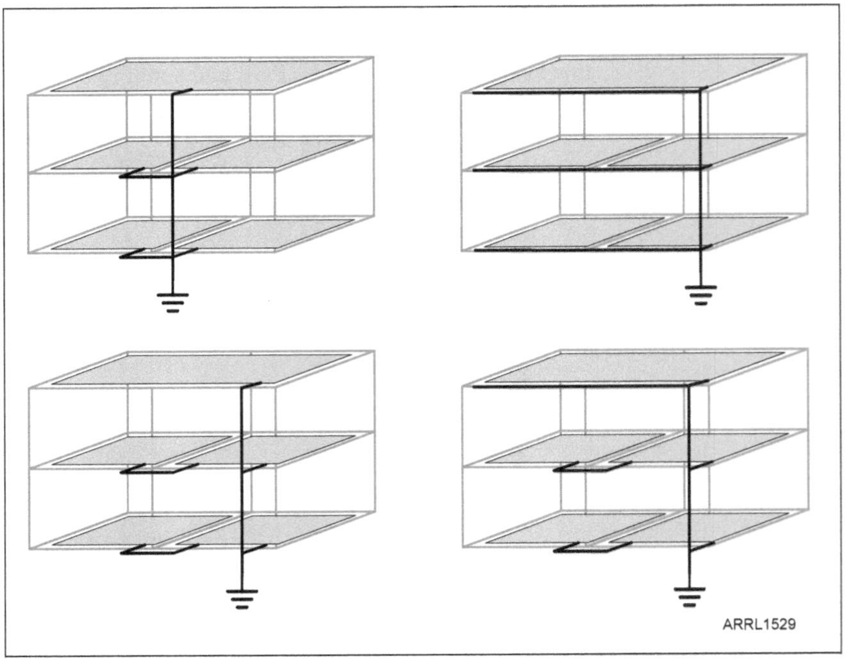

Figure 6.16 — For equipment on shelves, create a ground plane on each shelf. Bond each ground plane segment together (heavy lines) and route cables between the segments along the bonding strap or wire. Several possible examples are shown to give you an idea of how to connect ground planes together. Most station arrangements will be considerably simpler.

board and secured with clamps, or held down by the weight of the equipment on it. The metal surface should extend under the equipment as much as practical. It is not necessary to cover the entire operating surface — strips that are a few inches wide will suffice. The important thing is to get a lot of continuous metal surface close to the equipment and interconnecting cables.

If your equipment is on wooden or plastic shelves, create a ground plane on each shelf and bond the individual ground planes together. **Figure 6.16** shows some possible arrangements. Metal shelves can serve as the ground plane themselves. Keep the plane-to-plane connections short and use wide strap or heavy wire. Then tie the connection into the rest of your ground system. For connections between equipment on different shelves, run the connecting cables along the bonding conductors that connect the ground planes together.

RF Bonding Bus

The bonding bus can be any wide conductor, including strap, angle, or tubing. Copper and aluminum are both excellent materials for this use. Connections can be made with screws, terminals, or clamps intended for wire connections. If the ground plane material is heavy enough, you can make grounding connections directly to it without a separate bonding bus. As discussed previously, use bonding jumpers of heavy wire or strap or flat-weave braid. Ground terminals or enclosure mounting screws with a clean metal connection (no paint or insulation) are convenient places to attach the jumpers.

The Power Supply Negative Return

If a dc power supply's negative or "V–" terminal is not connected to its chassis or enclosure, the supply is called "floating." This is common for laboratory-style or modular supplies designed to be built-in to other equipment. If a floating supply is used for powering station equipment, connect the negative output directly to the largest load's ground terminal — usually on the transceiver.

For supplies with the negative terminal connected to the enclosure, be wary of resistance in the negative power conductor. The resistance can create a voltage drop that causes what sounds like "RF feedback" and other distortion under heavy loads, such as when transmitting. Don't assume the negative terminal is well-connected to the enclosure. Verify that there is a good connection with a voltmeter. Paint on the enclosure or a loose bonding jumper from the V– terminal can create a poor or nonexistent connection.

Whether or not the supply is floating, the enclosure should be connected to the ac safety ground by the power cord connection or a bonding jumper. For the power connection to station equipment, be sure to use heavy wire and make good connections. The power supply enclosure should also be bonded to the station ground system. For a more complete discussion of these issues, see the link titled "Power Supply Bonding and Audio Distortion" at "**http://k9yc.com/publish.htm**" **k9yc.com/publish.htm** for K9YC's 2009 paper titled "Power Supply V-Bonding, Hum, Buzz, and RFI."

Make it easy to connect short jumpers from each piece of equipment to the bonding bus. Figure 4.19 in the chapter on Lightning Protection shows a typical example of how multiple pieces of equipment might be connected to a bonding bus, with or without a ground plane. In the case of Figure 5.5, sheet metal screws were installed every few inches for ring or fork terminals. As long as the connection is direct and robust, feel free to improvise and substitute materials you have. If ½-inch copper pipe is used, acorn clamps can be used, as well.

If it is difficult to get access to the back of the equipment when it is in the operating position, try to create ground connections that can be attached as a plug-and-jack without access to a nut or screw. For example, banana plugs and jacks are easy to connect "by feel" and will stay in place once inserted. The goal is to have short connections without loops or bends.

Figure 6.17 — A typical mini-desktop computer with a short bonding jumper from a clean metal-to-metal enclosure screw to the bonding bus.

Computers

Computers can be quite susceptible to RF-related problems, so it is important to bond them to the ground plane and bonding bus. **Figure 6.17** shows a typical mini-desktop PC enclosure connected to the bonding bus with a short piece of heavy wire. There are several enclosure screws available, but this was the closest one with a clean metal-to-metal connection between the main chassis and the outer enclosure.

Larger desktops may have to sit on the floor or a shelf away from the ground plane — use the shortest practical strap to connect them to the bonding bus. Bundle the cables together using hook-and-loop strips, wire ties, twist ties, or similar products. A ferrite RF-suppression clamp-on core with multiple turns may be necessary on signal cables. Route the cables along the bonding strap and route both directly to the ground plane.

Laptop computers are not designed to be connected to a ground system. However, if they are connected to the radio equipment by audio or control connections, a bonding conductor should be added. The laptop "chassis" connection is generally available at the locking screw terminals of a D-style 15-pin or DVI video connector. Several D-type connectors with suitable threaded inserts are shown in Figure 6.17 above the ground jumper attachment point. HDMI and USB connectors do not generally have an accessible ground terminal, although their metal shells should be connected to the chassis or ac safety ground.

Most RFI problems with laptops are the result of RF being picked up by connecting cables and can be addressed with ferrite cores. Touchpad pointing devices can be sensitive to RF, as well. A standalone mouse or trackball with ferrite cores on the cable at the laptop are often a solution. Inexpensive peripherals, however, may not be capable of reliable operation in the presence of strong RF fields, even with extensive preventive measures. The best solution in these cases is to replace them with higher-quality devices. Your club members may be able to recommend suitable replacements.

Computer cables can be difficult to manage at RF. They are often much longer than necessary for the specific connection, leaving lots of exposed cable to pick up RF. Since the connectors are specialized and usually molded to the cable, they cannot be cut to length.

Many cables include molded-on type 43 cores installed to meet FCC emission limits at VHF and UHF with little suppression at HF. Multi-turn ferrite chokes of type 31 material can be used to deal with RFI at HF (see K9YC's tutorial referenced earlier). In addition to the ferrite core, coiling the cable and laying it directly on the ground plane as in **Figure 6.18A** will reduce RF pickup. Wire ties, twist-ties, or tape will keep the cable coiled.

Should You Disconnect?

Disconnecting antennas feed lines is frequently suggested for stations that do not have a good grounding and bonding system. However, as the experts point out, unless you disconnect *everything* going in and out of the station, disconnecting feed lines is unlikely to be as effective as lightning protectors on an SPGP and good bonding of equipment. If you decide to disconnect the feed and control lines, and they are normally connected to lightning protectors, be sure to disconnect them from the *equipment* side (not the *antenna* side) so the protector is still connected to the antenna. If no protectors are being used, disconnect the cables *outside* the building and keep them several feet from anything that might offer a path inside. Inside the station, once a lightning surge is inside the zone of protection, equipment protection depends on your bonding scheme. You may wish to short the RF inputs to ground manually, with a suitable antenna switch or antenna protector device. Opening the antenna connection with a switch, however, will offer no protection against lightning's spark gap of thousands of feet.

(A)

(B)

Figure 6.18 — Computer cables, such as the USB cables in A, can be coiled up and laid directly on the RF ground plane to help minimize the amount of RF picked up. At B, the completed station is shown. Notice how all of the equipment is completely or partially on the RF ground plane.

6.5 Practical Stations

Regardless of whether the station is large or small, permanent or temporary, in a basement or apartment, every station builder needs to answer these basic grounding and bonding questions:

• Are all exposed metal enclosures and connectors in my station connected to an ac safety ground?

• Is every path in and out of my "protected zone" protected?

• Are all of my ground electrodes bonded together?

• Are my external antenna supports, antennas, and accessories grounded properly?

• Is my equipment bonded together properly for ac safety, lightning protection, and RF management?

• Have I minimized interconnecting cable length and loop area?

Use the set of tools and techniques in previous chapters to apply the recommended grounding and bonding practices in your station as best you can. If you follow those guidelines in addressing these questions, your station should be electrically safe, resistant to damage from lightning, and experience a minimum amount of RFI. When problems do occur, you have the necessary information to troubleshoot and maintain the station, as well.

Don't be afraid to try different arrangements of equipment and grounding and bonding. As long as the basic safety requirements are met, there are lots of different ways to create an effective station. When you buy a new piece of equipment, or add a new antenna, it is a good time to review your station and see if there isn't a better way to do things. Most hams rebuild their station from time to time, to improve performance or expand capabilities. That's a great time to improve your grounding and bonding practices, too.

As mentioned earlier, you might not want or be able to tackle all aspects of grounding and bonding at the same time. It can be a big job! You may also have equipment already installed and wired. Taking apart the whole station might not be something you want to do either. The following list provides some suggestions for taking things one step at a time. Start by answering all of the questions above so you know where you need to focus your efforts. By focusing on one type of work, you'll have all the tools and materials on-hand at the same time which makes the job easier and results better.

Step One — AC Safety: Do a thorough check of all your residence's wiring and correct any problems or weaknesses. Verify that every piece of equipment in your station has a secure ground connection. If there is any electrical work to do at your ac service entry panel, get that done (or at least inspected) by an electrician.

Step Two — Comprehensive Lightning Protection Plan: Develop your plan first, even if you aren't going to do everything in the plan right away. If you have a plan and follow it, each step will improve your station and you will avoid having to rework something later.

Step Three — AC Power Distribution: Upgrade your station's ac power wiring, if necessary, with new branch circuits, distribution strips or centers, a master power switch, and so forth. Plan for integrating your ac power and bonding systems.

Step Four — Ground Electrodes and SPGP: Install additional ground rods or whatever is appropriate for your station, including bonding to ac power system grounds. Build or improve ground systems for towers, masts, and antennas, including bonding to other ground electrodes. Create and install an SPGP.

Step Five — Bonding Equipment and RF Management: Install or upgrade your equipment bonding system. Bonding conductors should all be strap or heavy wire as necessary. Either bond equipment directly together or install a bonding bus and ground plane. Tie the bonding system securely to the ac safety ground conductor and SPGP.

First-Floor and Basement Stations

If your station is located in a residence on the ground floor or in the basement near ground level, following the guidelines is straightforward, if not always easy. Start by mapping out where your ac power will come from and where you have access to external walls. Review your lightning protection plan, and plan your external ground electrodes. Some helpful suggestions are:

- Locate your station in a room with an exterior wall, if possible.
- Minimize the length of connections to external ground electrodes.
- Basement stations should minimize the possibility of shock through a concrete floor or walls.

Upper-Floor Stations

Stations on a second floor or higher pose a challenge to station build-

Getting Behind Your Equipment

Speaking from the experience of many hours spent on hands and knees, leave room to work comfortably behind your station, if possible. If you don't have room to leave space behind the station permanently, make it easy to pull the equipment table or desk away from the wall temporarily. Having clear access makes it so much easier to route power cords and connecting cables cleanly and to keep bonding conductors short and direct with good connections.

Window-Mounted Feed-Through Panels

Slim metal panels that fit into windows and eave openings with multiple feed-through connectors are convenient ways to get feed lines and control cables into a building.

Many feed-through panels have an opening for control cables that bypass any protection and would defeat any protective function of protectors mounted on the panel for other cables. Protectors must be mounted on the SPGP. If the panel features a width adjustment, that also limits good contact between the sections. It is not recommended that unprotected open-wire feed lines be brought into a building through these panels. Open-wire lines should be protected outside the building. Be wary of installing a feed-through panel near metal flashing, roof components, or window frames since lightning could easily jump to them from the feed lines and the panel.

If you use such a panel, do not rely on it as a SPGP. It should be bonded to the SPGP with strap or heavy wire as if it were another equipment chassis. It should not be in series with the SPGP external ground connection — run a separate wire or strap from the panel to the external ground system.

Most of these panels are too small for protector mounting (except for possibly through-panel mount protectors, such as the PolyPhaser B-series), and especially for mounting ac power and control line surge protectors. Any protectors built-in or mounted on a feed-through panel should be considered as secondary to those on the SPGP. If the SPGP is close to a window-type feed-through panel, additional protectors are not necessary or recommended.

ers, particularly regarding lightning protection. These stations are more like airplanes than ground-mounted equipment. As in an airplane, bonding to minimize voltage differences takes a primary role in protecting equipment from lightning and minimizing problems caused by RF being picked up on the station feed lines and wiring.

AC Safety Grounding

The required connection to an ac safety ground can still be made through the power wiring at a wall outlet. Remember that the distance to the ac service panel and any building ground electrode will be long and of almost no effectiveness for RF or lightning protection. In some buildings, including houses, the upper floor(s) may have an ac power distribution panel for that floor. It is best to confirm that the safety grounds of the outlets on that floor are tied into the building ground bus and ultimately, into the main power panel for the building. If the building is steel frame construction, the frame will be connected to the ac safety ground. Check with local building codes regarding use of the frame as a ground connection.

Lightning Protection

The connection to an external ground system depends on the height of the station and is approximately 10 feet per floor. The inductance of such a connection is too long to be effective as a lightning ground. On higher floors, an external lightning protection ground may be required to meet code requirements but is unlikely to provide much protection to your station equipment. This is often a difficult situation, and no suitable connection may be available. However, the creation of an effective SPGP and bonding equipment together will help prevent equipment damage. At RF, however, the station will behave as if it is ungrounded. The biggest challenge is in minimizing the inductance of the ground connection. Possibilities for the ground conductor include a wide copper strap (6 to 18 inch, 16 gauge minimum), a copper downspout, or higher-resistance galvanized conduit (EMT). There are few other choices but any connection to the external ground system should be as direct as is practical.

In a steel-frame building, the frame may serve as the ground conductor. A railing or door or window frame may be connected to the building's steel frame. A water pipe or radiator may also provide the necessary connection. Before counting on a building frame connection as a bonding path for lightning protection, use a multimeter to verify that it has dc resistance of 1 Ω or less to the ac safety ground in an ac outlet.

Figure 6.19 shows two common methods of routing feed lines and control cables to an upper floor station, either from ground level or above ground directly to the station. In both cases, an SPGP must be created at the station as close as practical to the cable entry point with its own grounding conductor to the external ground electrode. All lightning protectors must be mounted on the SPGP so they will fire at the same time.

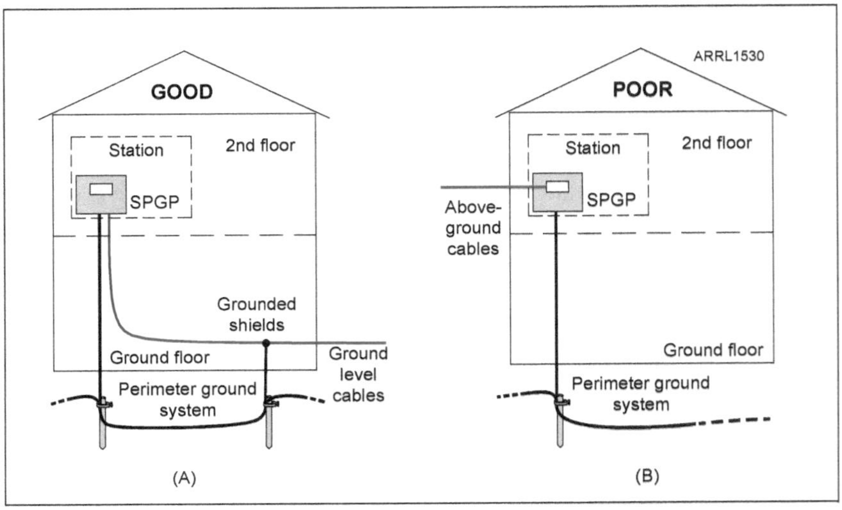

Figure 6.19 — Two options for lightning protection with feed lines and control cables entering an upper-floor station. The connection from the SPGP to the external ground system should be large strap or metal pipe (see text).

Even if lightning protectors are added elsewhere outside the building, an SPGP at the station is still required. If a separate feed-through panel is used, it must be bonded to the SPGP. The SPGP must have its own grounding conductor to the external ground electrode.

There are many installations in which ham radio antennas are mounted on the roof using a pipe mount. Providing a viable path for lightning energy that couples to such antennas can be a challenge. Even though the path down to an external earth ground will likely be long, it is still desirable (and often mandated by code) to ground those roof antennas. If the station is located on an upper floor under that area of the roof, the protection challenges are even greater. Remember that all ground rods must be bonded together as part of the perimeter ground system!

Whether the cables are run to the station from ground level or above ground, all cables must be run outside the building until they reach the entry point where the SPGP is located. Your first choice should be running above-ground cables to ground level before they reach the station entry point. At ground level, cable shields should be bonded to the external ground system using a cable grounding kit or bonding bracket. If a pipe or downspout is used as an SPGP ground conductor, the cables can be run through it as a cable protector. The inductance of the long ground connection will exaggerate the strike voltage build-up on the SPGP and connected equipment because of the ac components of the strike energy. This higher voltage makes the area more vulnerable to flash-over (arcs between equipment, wiring, and metal surfaces). Even though it may have high inductance, the ground conductor and ac safety conductor still provide a dc connection and will allow the SPGP and the connected equipment to return to near ground voltage over several hundred milliseconds.

If the antenna cables are to be run through the house for a considerable distance, to minimize the surge energy on those cables for safety reasons a secondary SPGP without protectors can be established and connected to the perimeter ground near where the cables enter the structure. The creation of a secondary SPGP and the safe routing of the coaxial cables through the house structure is beyond the scope of this book — a lightning professional should be consulted.

If the SPGP has been constructed properly and equipment is well-bonded, there is no need to disconnect. If you must disconnect, do it from the "Equipment" side of the protector and not the "Antenna" side so that the surge protection device is still connected to the feed line. Regardless of which method you use, route unprotected antenna cables well away from protected cables. Routing the protected and unprotected cables in a single bundle may look neater but allows lightning to jump around your protectors with ease.

RF Management

RF management in an upper-floor station is handled in the same way as for ground-floor stations. The RF fields from your antennas will probably be stronger than at ground level, requiring more attention to equipment and ground plane bonding. Your RF bonding system must connect to the ac safety ground as always. If the only access to the ac safety ground is via the center screw of an ac duplex outlet, use it but be aware that RF management is still likely to be a problem.

If operating on multiple HF bands produces RF hot spots on your equipment, try placing lengths of wire along the wall or floor to act as a counterpoise RF ground. Another option is to use an artificial RF ground as discussed in the RF Management chapter. See also the sidebar "Dealing with RF Hot Spots" later in this chapter.

Towers Next to Building

If a tower or mast is mounted close to the building as in **Figure 6.20**, it must be well-grounded to create a dominant path to the earth. The ground system must include buried radials to dissipate charge away from

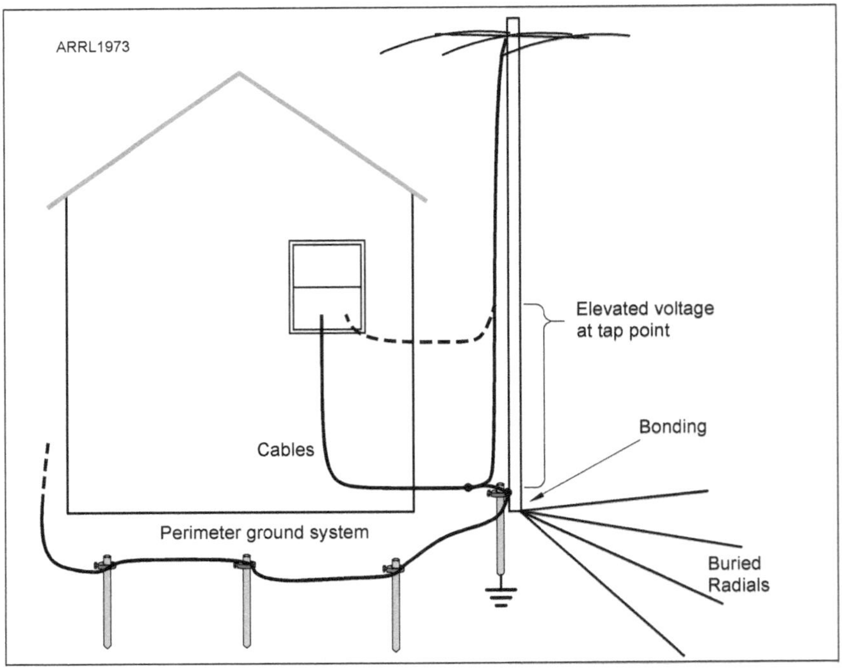

Figure 6.20 — Bonding of cables and ground systems for towers next to buildings. Feed lines and other cables must run all the way to the ground to minimize voltage on the cables created by a lightning surge.

the building. It must also be bonded to the perimeter ground system and all external ground electrodes. The goal is to shunt as much lightning strike energy to the earth as possible.

Run feed lines and control cables all the way to the bottom of the tower and bond them to the tower at the bottom. If the cables are routed away from the tower above the base, during a lightning strike they will be elevated to a voltage roughly equal to the fraction of tower height times the voltage at the top of the tower. See the previous section for a discussion of the station SPGP.

Wire Antennas and Ground-Mounted Vertical Antennas

This book focuses on tower- or mast-mounted antennas because they are exposed and grounded. Many smaller HF-oriented stations use wire or ground-mounted vertical antennas instead.

For wire antennas, all feed lines are routed to an SPGP whether they are coaxial cable or open-wire line. (See the Lightning Protection chapter for a discussion of open-wire line protection techniques.) If possible, route the cables so they are not in contact with the building.

For ground-mounted verticals, even if a ground screen of radial wires is used as a counterpoise or the antenna is a "ground independent" design, a connection at the base of the antenna to a ground rod or other electrode is required for lightning protection. Use heavy wire or strap to make the connection. This "extra" ground connection will not adversely affect the antenna's performance. Don't connect a ground conductor directly to the antenna radiating element. The manufacturer of a commercial antenna will provide instructions on how to make the connection. For homemade verticals, a grounding bracket, similar to those shown in Figure 4.9, can provide the connection point for the bonding jumper between the ground rod and the antenna. The lightning protection ground rod must be connected to the other parts of the ground system. Do not use lightning protectors at the antenna — they should only be mounted on the station SPGP.

Temporary, Portable, and Field Day Stations

You will find that setting up an effective station at ARRL Field Day or for a public service event (including emergency communications) presents a variety of RF and safety challenges. The challenges appear different every time, but separating them into our three categories of ac safety, lightning, and RF will help you avoid problems.

AC Safety Grounding

If you are operating from commercial mains, make a visual check and use an ac outlet tester to be sure that power is wired correctly at the source and that the voltage is within range. Don't assume that the outlets are in good shape and wired correctly. If you don't think the ac supply is safe or if the voltage isn't right — don't use it! Be wary of using ac from different outlets at the same station because they may be on different branch circuits that use different neutral and ground conductors, leading to possible shock hazards along with hum and buzz issues. A building engineer or maintenance supervisor may be able to help with problems and identify good sources of power and a good ground connection. Safety is not optional!

Generator and battery inverter power pose their own problems. As the chapter on AC Power System Grounding and Bonding points out, you are not required to use a ground rod if the generator outlet grounds are bonded to the generator frame. Ground rods may be used, however, if desired. If the generator is more than a short distance from the station or if more than one separate station is powered by the same generator, use a ground rod at the generator and at each station with three-wire power cords. If more than one generator or inverter is used, either bond their grounds together or don't connect more than one to the same station.

As with commercial mains, inspect the generator and the outlet wiring! Don't use the generator if it's not wired correctly or if the voltage is out of spec. If you are planning such an operation, check out your generators and inverters well in advance.

Each station should be equipped with:

- AC voltage meter — don't rely on the brightness of a light bulb!
- GFCI-protected outlets (see the sidebar "GFCI in a Box," earlier in this chapter)
- Overvoltage protector (see *The ARRL Handbook* chapter on Power Sources for suitable designs)

RFI from inverter generators has been reported. If this is a problem, use a power-line ferrite choke as described in recent editions of *The ARRL Handbook* in the chapter on Assembling a Station.

Lightning Protection

It is not reasonable to expect a temporary installation to provide the level of lightning protection that can be created for a permanent station. You should be prepared to disconnect or lower antennas if there is any chance of lightning occurring in the vicinity. If antenna cables are disconnected, place their ends well away from personnel and equipment. A miniature SPGP can also be created by mounting a lightning protector and an ac power protector on a single ground rod at the station.

A lightning detector, such those available from Weather Shack (**www.weathershack.com**) or other vendors, is a good addition to your equipment. Lightning monitoring apps for smartphones, tablets, and PCs are widely available. If lightning is in your area — which can be a radius of up to 10 miles — get off the air and protect yourself. Be safe!

Portable stations pose their own risks, since they are often operated on hilltops or in clear, exposed locations — hams love high places for operating! Summits On The Air (SOTA) veteran Steve Galchutt, WGØAT, and his pack goats have operated from many a high place. I asked him about lightning safety when operating on the trail. "If dark clouds are in the area I pack up and without delay head for cover. While SOTAteering, once I hear that familiar snap-crackle-pop in my receiver, I hurry up and finish the pileup (sometimes QRTing immediately!) and begin to hasten my tear down to get ready to head down to lower elevations! Mother Nature rules! She will beat you! Pay her respect and you will keep breathing!"

Steve notes there are inexpensive lightning detectors, "but even an old AM transistor radio tuned between stations to listen to static will act as a detector. It takes some practice but once I hear pop-pop about 2 – 3 seconds apart I go on alert and begin to think it's time (to get to a safe location)." Clear-sky lightning is not uncommon at high elevations.

"Mountain wisdom says 'stay off the peaks after 12 noon or sooner if dark clouds appear threatening!' and you'll live another day!" The same goes for Field Day, hilltopping, and mobile operation.

RF Management

Assuming you are setting up a station from individual pieces of equipment, treat it as a standard desktop station and create an RF ground plane under the equipment or a heavy bonding bus at the rear of the station. A metal folding table works particularly well in this situation, both electrically and mechanically. A roll of flashing strip, metal screen, or even aluminum foil can be spread out under the equipment. Whatever is used as the ground plane, use short jumpers with heavy alligator clips to

Dealing with RF "Hot Spots"

The chapter on RF Management discusses using a resonant, quarter-wavelength wire to create a low-impedance point in a station. This technique can come in very handy in portable or temporary operation where the station configuration tends to change every time and creating a full-blown ground system is impractical.

Experience will eventually teach you which bands tend to have problems with the equipment and antennas you use. Cut and tune quarter-wavelength wires for those bands, insulate one end and attach the other to an alligator clip, and bring them with the station. If a "hot spot" appears, attach the wire for that band at the point of the hot spot. Keep the open end of the wire where it can't be touched or stepped on. Even QRP operation is sufficient to create enough voltage at the open end for a tingle and the voltage from 100 W operation can actually set grass or leaves on fire! So be careful where you run the detuning wire.

connect the equipment to the ground plane or bonding bus, which is then jumpered to an ac safety ground point.

If a ground plane is impractical, bonding equipment enclosures directly together is a good alternative for a temporary setup. The bonding jumpers should use the same heavy wire used for a home station. If a Powerpole or quick-disconnect connector is used on the "free" end of each jumper, they can be connected however the station is configured. One of the bonding jumper connection points is then used as the connection point for a jumper to the station ground system. **Figure 6.21** shows the general idea. This is not recommended for permanent installations because lightning surge current will flow between equipment instead of through a bonding bus or ground plane.

Portable stations can range from collections of loose gear assembled and disassembled at each location to "stations in a box" that are mounted in a carrying case semi-permanently. Usually battery or vehicle powered, portable stations are free of ac safety grounding issues, but RF management can still be troublesome. It is a good idea to have a ground stud or terminal available on the rack frame for connection to an external or vehicle ground if needed. For popular QRP-style operation with miniature transceivers, antenna tuners, and keyers or keyboards, some replaceable aluminum foil or metal screen can serve as a lightweight ground plane. A piece of heavy bare wire can serve as an RF bonding bus. A set of flexible jumpers with alligator clips that can act as temporary bonding jumpers, too.

The author has built a small station that can be installed in a car under a passenger seat or removed to operate from a table. (See **Figure 6.22A**.) A typical all-band 100 W mobile transceiver plus an automatic antenna tuner, screwdriver-style antenna control, and power distribution strip are all attached to a heavy aluminum clipboard with the clip removed. (The

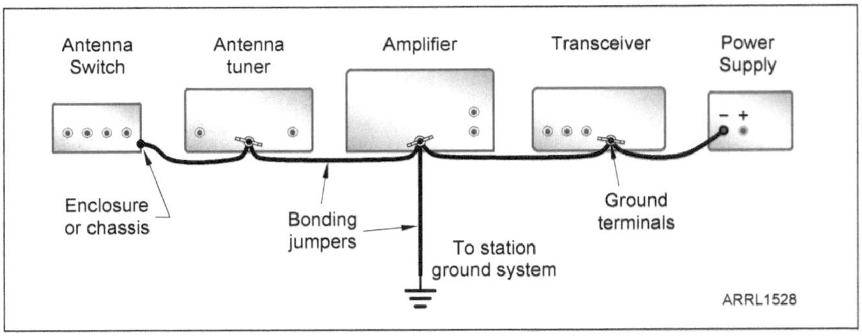

Figure 6.21 — Direct equipment-to-equipment bonding. Use heavy wire or strap connected to a single ground terminal or enclosure screw on each piece of equipment, including computers and related gear (not shown).

(A)

(B)

Figure 6.22 — A portable station with a transceiver, antenna tuner, SWR sensing unit, and dc power distribution strip (A). The equipment is mounted on an aluminum clipboard using heavy hook-and-loop strap. All equipment is bonded to the clipboard with heavy wire jumpers. If installed in a vehicle, a ground jumper from the vehicle frame can be attached. The entire package can be removed from the vehicle for portable operation, as well. Portable audio racks (B) with one or more metal shelves also make good self-contained stations for portable and mobile use.

transceiver's front panel is removable and connected with a separation cable.) The equipment enclosures are all jumpered to the metal clipboard, which serves as the ground plane and bonding bus. It is easy to attach or remove ground jumpers for other uses of the station. The sturdy clipboard serves as both a ground plane and a mounting plate.

Another common approach is to build "mini-rack" stations for emergency communications and public service use as shown in Figure 6.22B. Mobile and portable stations, such as "rover" stations that operate on many VHF/UHF/microwave bands during VHF+ contests, also mount many pieces of equipment in portable rack-style cabinets or containers.

Mobile Stations

Vehicle Construction

Modern vehicles cannot be considered *all-metal* because composites and other non-metallic materials have been used in increasing amounts in their manufacture. Since materials can include combinations of metal and non-metallic materials, we'll use *composites* as a universal term, except in cases where the surface is well-known, such as the now-common use of aluminum for outer body panels.

The once-standard body-on-frame construction has largely given way to *unibody* construction, even for heavy-duty trucks. Hoods, fenders, trunk lids, truck beds, and other body panels are often made from composites. Most composites used in common vehicles have an underlying steel framework. This is important to consider, especially when mounting mobile antennas.

You can use a magnet to determine if a material is ferrous (steel). However, some composites (steel and rubber sheet combinations, for example) will still attract a magnet, but to a lesser degree than heavier steel

Warranty Issues

No discussion on installing mobile equipment would be complete without noting possible warranty issues involved. Check your vehicle's warranty coverage carefully! Ford Motor Company for example, warrants their aluminum-bodied vehicles for five years, regardless of mileage. However, if customer actions and/or extreme environmental conditions cause perforation of the body's sheet metal, then there is no warranty coverage! To accommodate installation of extra equipment, all vehicle manufacturers have recommended procedures for installing aftermarket devices, including amateur radio gear. These recommendations are readily available on line, or from your dealer in question, and often include toll-free telephone numbers. All of this information is included in your vehicle's owner's manual.

panels. Mounting antennas atop any composite, regardless of its construction, is ill-advised due to questionable strength. This includes so-called groundless half-wave antennas, which tend to have excessive common-mode current on the feed line without a choke or at least some conductive layer directly under them to decouple the antenna and feed line. (See the Appendix for choke designs.)

Composite-bodied vehicles often have an underlying steel framework. This framework may not offer a dc or RF path to all of the vehicle's various body components. A dc path can easily be verified with the use of a DVM (digital volt-ohm meter). An RF ground plane is questionable if it isn't all-metal and directly under the base of the antenna!

DC Power

If you look closely at low-current factory wiring, you'll notice that specific circuit subsystems, such as the headlight subsystem, use a common chassis ground point that is sometimes referred to as a *star ground.* (Star grounds were also discussed in the chapter on Lightning Protection.) Other subsystems may use a different ground point. These ground points provide a common voltage reference for that circuit subsystem and are typically not used by more than one subsystem. Do not use them as radio equipment power return grounds!

As current flow changes in the circuit subsystems, a voltage difference can be created between the subsystem common ground points. These voltage differences can mimic RFI, which makes finding and eliminating them difficult. To avoid these voltage differences, *home-run wiring* (separate return wiring that does not use the vehicle body as a current path) is an important power wiring technique for amateur radio gear. It can also minimize dissimilar-metal, galvanic corrosion in aluminum-body vehicles (discussed below).

High-current applications, such as starter motors and battery charging, do not use the steel body or frame structure as a ground return. Separate home-run wiring to the battery is used. The home-run returns are connected at, or very close to, the battery's chassis ground. This avoids interference with the vehicle's *Battery Monitoring System* (*BMS*). **Figure 6.23** shows the basic structure of where to connect radio equipment in non-electric vehicle power wiring. An electronic current sensor is mounted directly at the battery's negative terminal. If your vehicle has a battery current sensor, do not connect negative or return wires directly to the battery's negative terminal, which would bypass the sensor and upset the BMS calculations.

The BMS accounts for all charging and discharging battery current. This gives the vehicle control systems an idea of battery health and how to

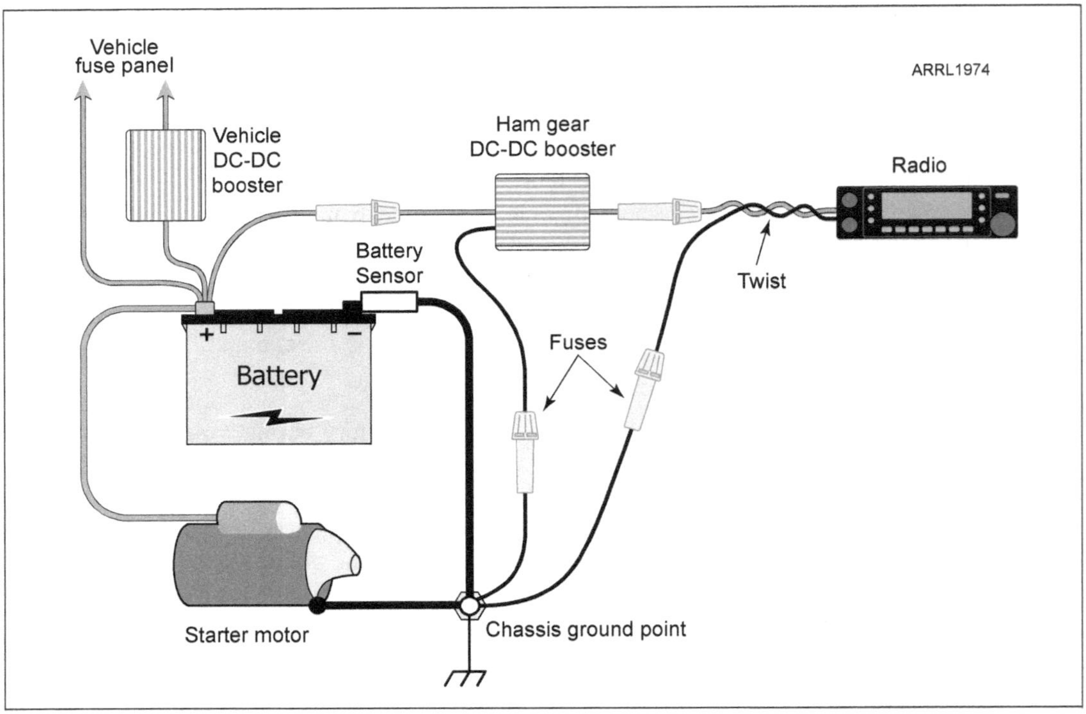

Figure 6.23 — Basic connections of amateur equipment to a vehicle electrical system. Use twisted-pair cables for powering the radio and any dc-dc boosters.

make use of battery power. This is particularly important for vehicles with *Engine Idle Shutoff* (*EIS*) that require frequent use of the battery to restart the engine. Vehicles with EIS typically incorporate a *dc-to-dc boost* or *battery boost* module. Whether powered by a separate battery or super-capacitor, the boost modules provide a steady dc voltage to lighting, entertainment units, and other systems during engine shutoff and starting. These boosters *should not be used* to power amateur radio gear!

It is common to use a dedicated battery booster for powering radio equipment so that the supplied voltage is always 13.8 V (or higher) regardless of the battery's charge level. (The booster must have a low-voltage cutoff threshold to prevent the battery from becoming too discharged to start the vehicle.) The battery booster input and output dc wiring should be treated the same as if the wiring was connected directly to the battery. Use heavy, twisted-pair wiring with fuses in each line and connected to the battery's positive terminal and to the battery's chassis ground point.

Conductors of the heavy zip-cord style power cables supplied with the transceiver should be twisted together (or replaced with heavy twisted-pair wiring) as described in the RF Management chapter, with the positive wire directly to the battery and the negative or dc return wire attached to the chassis ground point as shown in Figure 6.23. An RF choke installed at the transceiver is also recommended to further limit RF coupling to the power wiring. (#10 AWG paired wires are suggested — see the Appendix.)

DC power cables provided with transceivers often have to be lengthened from their stock length to reach a vehicle's battery. To ensure that the additional resistance does not result in excessive voltage drop, the complete cable (from power source to transceiver) must be increased in size. If the length of the cable is doubled, the cross-section area of the power wiring must be doubled (or the resistance per 1000 feet halved) to present the same resistance. For example, if a #12 AWG power cable is increased from 6 to 12 feet long, the next largest size wire with at least twice the cross-section area is #9 AWG. (#8 AWG is more common and probably less expensive than the less-common #9 AWG.) A table of wire sizes is provided in the Appendix as Table A.1.

Bonding in Vehicles

While plastics and composites are increasing common, metal is still used in the underlying supporting structure, especially for the chassis and suspension. When connecting to a metal structural component, use an external-tooth star lock-washer and an anti-corrosion compound containing gluconates between the bonding material (usually braided or solid copper straps) and the ground point. Surface coatings should not be removed. Star washers are designed to cut through this layer without causing oxidation. Solid strap is preferable, but flat-tinned braid may be needed due to vibration. (See **Figure 6.24.**) Bare braid will corrode in this application, making it ineffective at RF although it may still be suitable for dc and low-frequency bonding. Use braid with a weather resistant jacket or covered by heat-shrink tubing, if possible. Do not use brake and fuel line brackets as attachment points, for safety reasons.

Mobile Station "Grounds"

Amateurs deal with a variety of vehicle "grounds": dc ground (power return); "RF ground" (local RF reference voltage — see the Grounding and Bonding Basics chapter); and ground plane (surface for RF signal flow to an antenna feed point). Any point represented as "ground" can be intended to perform any of those functions, so it is important to understand which function is being discussed. (Obviously, a mobile station will not have an earth connection!)

(A) (B)

Figure 6.24 — A vehicle bonding attachment point (A) and a typical exhaust system bonding braid (B).

Adding bonding between body sections, such as the bed and cab of a pickup truck, generally isn't necessary for supplying transceiver dc power or to control RFI. Bonding all sections together may improve antenna performance by creating a larger metal area to act as a ground plane. If the exhaust system is found to be radiating RF noise, a bonding strap or braid (see note above) can be attached to the piping with a stainless-steel muffler clamp or hose clamp as in **Figure 6.24B**.

Bonding Amateur Equipment

Amateurs with a home station are used to the idea of a metal surface being a relatively passive part of the station wiring that can be used as a bonding path. In a mobile station, however, the vehicle frame and body panels are an integral part of the antenna system, especially at HF. This makes bonding to the vehicle body a likely source of RF common-mode current.

In addition, vehicle wiring sometimes uses chassis ground points for vehicle electronics that may create significant currents and voltage drops between them. For this reason, using the vehicle body as a shield or bonding path at RF can lead to erratic performance and RFI, particularly to vehicle electronics, which could disrupt their function.

With few exceptions, the metal enclosure of amateur transceivers is connected directly to the negative power lead. Assuming the transceiver is correctly wired, no additional dc ground connections are necessary. The

Upfit Packages

Most automobile manufacturers offer what are called "Upfit Packages." These are special order items (both factory and dealer installed) that ease the installation of amateur radio gear. One common package is the factory installation of switched and un-switched, high-current wiring, and often a high-capacity battery and alternator. Another is a pre-installed, roof-mounted, NMO mount. These packages are often unknown except to dealer fleet salespersons but if you're ordering a new vehicle, inexpensive packages can be a real bargain and save much time and trouble. In addition, these packages have been designed for good radio performance and to minimize RFI.

only chassis connections for radio equipment should be the dc power return and the feed line shield at the antenna.

If the transceiver is the sole piece of equipment in the vehicle, connecting its enclosure to the vehicle frame is not electrically required. (It may be necessary to attach the transceiver to a vehicle body component to secure it for safety.) Connecting the transceiver's enclosure directly to the vehicle is likely to create an additional dc return path with the power return wire, which would provide a path for noise to be coupled onto the power connections.

Low-power accessories can use support structures for the dashboard, seats, and consoles as dc return connections if the main dc return path is not available. The same cautions about possible erratic performance and RFI apply, however. You can check for dc resistance and voltage drops with a multimeter.

It is common for a mobile station to consist of several pieces of equipment, such as a transceiver, antenna tuner, keyer, and antenna controller or switch. Bonding between pieces of equipment can present a challenge. With the equipment secured mechanically, bonding wires or strap can be "daisy chained" from enclosure to enclosure along with any control or RF wiring to minimize loop area. Another solution is to mount all of the equipment on a metal sub-panel and bond each piece to that panel before mounting the panel on a body panel. Finally, equipment can be mounted in a small metal rack or enclosure for audio or communication gear. The last two solutions work well for mounting equipment under a seat, in a trunk, or in a truck bed. (See Figure 6.22 and **Figure 6.25**.) Bonding the equipment enclosures, mounting panel, or rack to the vehicle is not recommended. Doing so may create RF paths that lead to erratic performance or RFI.

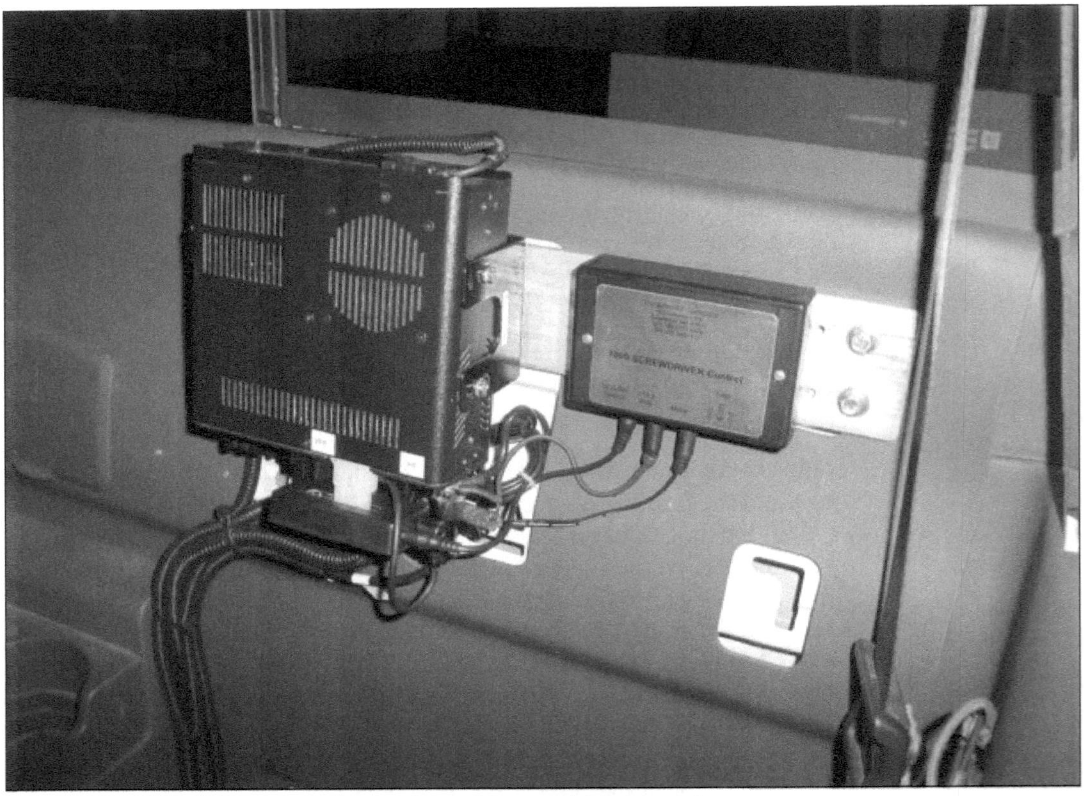

Figure 6.25 — Example of mounting equipment on a metal sub-panel.

When bonding amateur equipment, using heavy stranded wire or flat, tinned, braided strap are the most common conductors. These materials tolerate vibration well and allow equipment to be moved, if necessary. As mentioned elsewhere in this book, do not use unprotected braid removed from coaxial cable. Avoid using any type of braid for RF bonding where it is exposed to moisture, grease and oil, or corrosive chemicals. Ring terminals are preferred since they are less likely to vibrate loose than fork terminals. If the equipment is to be connected and disconnected frequently, Powerpole or quick-disconnect terminals should be used.

Transceivers often allow removal of the front panel to serve as a *control head* mounted away from the main radio by using a *separation kit* or *separation cable*. The control head is designed to be used without any ground connection. The only other connections that should be made to a control head are accessories such as the microphone, PTT switch, and a keyer or key/paddle. The control head should not be bonded to the vehicle.

Manufacturer Service Bulletins

Manufacturers have a lot of experience with installing radio equipment in their vehicles. This information is often available as service bulletins, which can be quite extensive. There may be a bulletin specifically for your vehicle, especially if it is sold in fleets for commercial or public safety use. These bulletins are usually available through a dealer's parts desk or service shop and may also be available online with a little searching of the company's website.

Some models have a different signal return in the control head than in the main radio leading to unpredictable operation if the control head returns are grounded. Follow the manufacturer's installation instructions to avoid problems caused by improper or unnecessary grounding.

Vehicle-Mounted Antennas

When mobile whips for HF through UHF are mounted above a conductive material such as a steel car body panel, the metal provides a path for RF current to flow to the feed point. This is referred to as a "ground plane" whether or not it is connected to the rest of the vehicle. Currents in the ground plane act as a replacement for the missing half of the dipole. To perform this function effectively, the conductive surface must be close to and coupled to the feed point of antenna and be electrically large enough to conduct sufficient current. (See the *ARRL Antenna Book* for a complete discussion of mobile antennas at HF and VHF/UHF.)

If the vehicle has a metal hood, roof, or trunk, it can serve as the antenna's ground plane at VHF/UHF. In trucks and some SUVs, mirror mounts or other external frames connected to the underlying metal structure may be sufficient. On the 21 MHz band and lower frequencies, such a mount is often insufficient. An effort must be made to use as much of the vehicle's metal frame and body panels as possible. This may require bonding body components together.

Wherever the antenna is attached, be sure the mount is not unintentionally prevented from making good contact with the body by a layer of paint or some other coating. Verify with an ohmmeter that the antenna mount body and feed line shield are solidly connected to the body panel. If no metal ground plane is available, the resulting RF current will flow on the outer surface of the coaxial feed line shield as common-mode current. This current will radiate a significant amount of RF that may cause RFI, including to the vehicle's electronics.

If no connections can be made to the metal body at the feed point, a bonding wire or strap can be attached between the feed point and the closest practical attachment point. Running a bonding wire or strap from a

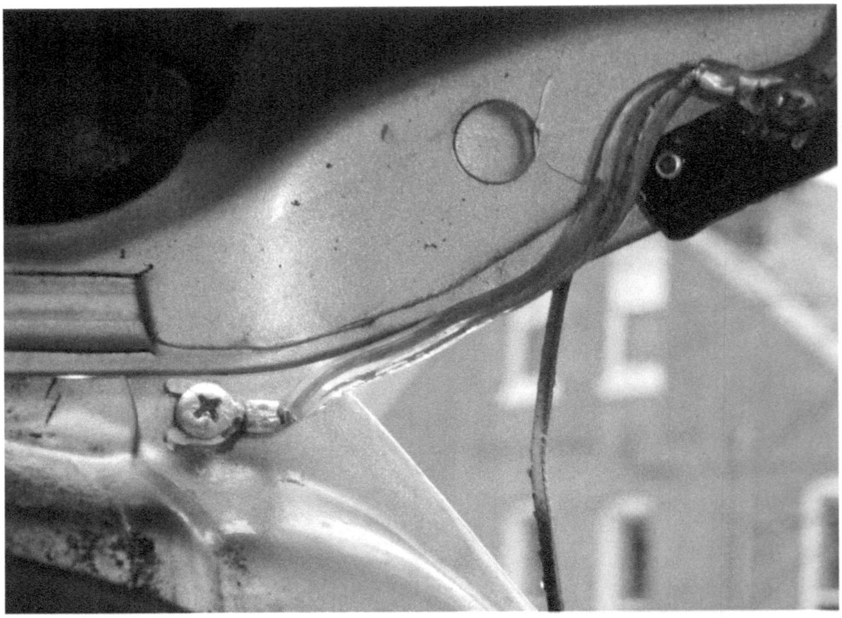

Figure 6.26 — Bonding an antenna mount to a body panel to make a better connection for the antenna's RF ground plane.

clamp-style mount to a nearby body screw is recommended. (See **Figure 6.26**.) Doors and hatchback hinges are frame-mounted and can provide attachment points for grounding conductors. When drilling a hole for a body attachment screw, be extremely careful not to drill into wiring that is often placed in hollow panels or frame members! Note that long straps or wires from the feed point to the vehicle frame are not a substitute for a ground plane and are actually part of the antenna.

Mag-mount antennas are intended to couple to the vehicle body through the capacitance of the mount which may be around 100 pF. This is enough capacitance to be effective at VHF (approximately 10 Ω of reactance at 2 meters) and UHF but not at HF, particularly the lower frequency bands. A three-magnet mount will work better at HF and usually has a terminal for attaching a bonding wire, as well. The bonding wire will become part of the antenna in parallel with the reactance of the mount.

Because they are exposed to vibration and corrosion, antenna mounts can also develop poor connections between the antenna and the mount, and between the mount and feed line. When this happens, the feed line's shield becomes the RF return path for the feed point, resulting in significant RF common-mode current. Check antenna mounts and their connection to the vehicle to be sure they are functioning as you expect.

RF Management and RFI in Vehicles

It is almost unavoidable in a mobile installation that the power/control wiring and feed lines of amateur equipment and the power/control wiring of the vehicle will have some level of common-mode RF current flowing on them. The usual cause is antennas mounted in such a way that allows a significant amount of RF common-mode current to flow on the feed line. HF mobile antennas with little or no metal mass directly below them will also couple significant amounts of RF to the feed line. Feed lines running along body components with poor or non-existent connections between them will also pick up RF current.

Another source of problems is from using non-proprietary microphones and other interconnecting wiring (such as for separating the control head from the radio) that have not been qualified by the radio's manufacturer. Use of an RF power amplifier will also increase RF common-mode current flow.

It is important to remember that the basic connections to power and to the antenna must be made properly. If temporarily grounding the enclosure of a single transceiver makes a significant change in an RFI issue, then some other ground connection, such as the dc power return or antenna mount, is likely to be missing, improperly connected, or inadequate.

The cabling for vehicle infotainment audio systems carries low-level analog and digital signals. This means RF common-mode current flowing on the system's interconnect wiring can cause RFI. RFI to safety-related sub-systems such as the anti-lock braking system (ABS) is also a serious concern.

Recognizing that it is better to fix the cause of RFI by reducing RF common-mode current, chokes should be installed on the coaxial cable at the antenna feed point. For equipment experiencing RFI, chokes should be installed at the equipment on any cables or wiring carrying the RF current that causes the RFI. Ferrite mix type 31 is the best material to use at HF through VHF and type 61 for low UHF. (See the RF Management chapter for more information.)

Vehicle wiring harnesses carry multiple cables, and it is often impossible to separate the individual cables. In addition, coupling between the parallel, closely packed cables makes using ferrite chokes on harnesses, even clamp-on varieties, difficult and often ineffective. The cables are, however, individually accessible close to the affected equipment where chokes can then be installed. (Bypass capacitors must not be used on digital signal wiring and analog audio connections.) Do not place a choke on a single conductor — that would degrade the signal it carries. Always include the signal return so that common-mode RF is blocked. If a single return conductor is used for multiple circuits (not good practice), all circuits and the return must be within a single choke.

Corrosion

Vehicle manufacturers zinc-dip steel bodies before paint is applied to reduce connection corrosion. Removing this zinc coating to achieve a good connection by sanding to bare metal *should not be done*. Use stainless steel hardware, galvanized star washers, and anti-corrosion products such as NO-OX for good surface-to-surface contact. Different techniques must be used on aluminum bodywork and with some composites.

Dissimilar-metal, galvanic corrosion occurs with any aluminum body structure unless certain precautions are used. Of particular interest is through-panel mounting of mobile antennas in aluminum-bodied vehicles. In these cases, separate home-run power return wiring should be used, regardless of the current level. Using the vehicle body as a power return can allow enough current to flow through the coax shield to cause dissimilar-metal, galvanic corrosion at the antenna's connection point, even if aluminum alloy mounts are used.

Bonding straps and dc ground returns used on aluminum-body vehicles must be attached using high-nickel content, stainless steel hardware, to lessen corrosion. Connections also require the use of anti-corrosive compounds containing gluconates.

Appendix

Resources for Information and Materials

The following is a list of websites and publications that you may find useful in developing and installing safety and lightning protection systems in your shack and your home. This is by no means complete, but provides a good sample of resources on the primary topics for this book.

There is no guarantee that website addresses will not change, and where possible, a stable top-level address is used. Should you find a website to be unavailable, the first step should be to do an internet search using the name of the site. This often locates sites that have been sold, renamed, or just moved to a different host.

Professional Associations and Companies

National Fire Protection Association (**www.nfpa.org**)

International Association of Electrical Inspectors (**www.iaei.org**)

Mike Holt Enterprises (**www.mikeholt.com**) — training and continuing education for electricians, many tutorials

PolyPhaser (**www.polyphaser.com/resources/white-papers**) — various papers and tutorials on lightning protection for communications facilities, including ham stations

Standards

National Electrical Code (NFPA 70) — available with a free profile at **www.nfpa.org/codes-and-standards/all-codes-and-standards/list-of-codes-and-standards/detail?code=70** (available from many libraries)

NEC Handbook 2020 — **catalog.nfpa.org/NFPA-70-National-Electrical-Code-NEC-Handbook-P16530.aspx** (available from many libraries)

FAA Document on Practices and Procedures for Lightning Protection, Grounding, Bonding, and Shielding Implementation — **www.faa.gov/documentLibrary/media/Order/6950.19A.pdf**

Motorola Publication R56 - Standards and Guidelines for Communication Sites — **www.blm.gov/sites/blm.gov/files/Lands_ROW_Motorola_R56_2005_manual.pdf** or **www.repeater-builder.com/antenna/site-stuff/are-fifty-six-man-2005.pdf**

IEEE Std 1100 – 2006 "IEEE Recommended Practices for Powering and Grounding Electronic Equipment" — **www.ieee.org** (available from many libraries)

MIL-HDBK-419A – Grounding, Bonding, and Shielding for Electronic Equipments and Facilities (Vol 1 and 2) —**ntrl.ntis.gov/NTRL/dashboard/searchResults/titleDetail/ADA239565.xhtml**

Books and Articles

Electrical Wiring

Cauldwell, R., *Wiring a House, 5th Edition*, The Taunton Press, 2014

Hartwell, Schwan, and Richter, *Wiring Simplified, 45th Edition*, Park Publishing, 2017

Laplante, P., Editor, *Comprehensive Dictionary of Electrical Engineering*, CRC Press, 1999

The Complete Guide to Wiring, 7th Edition, Cool Springs Press, 2017

Ugly's Electrical References—pocket-sized references of electrical information for wiring, common electrical circuits, and materials (**www.uglysbooks.com**)

Lightning Protection

Block, R. R., *The "Grounds" for Lightning and EMP Protection, Second Edition*, PolyPhaser Corporation, 1993.

Rand, K. A., *Lightning Protection and Grounding Solutions for Communications Sites*, PolyPhaser Corporation, 2000.

Uman, M. A., *All About Lightning*, Dover Publications, Inc, New York, 1986.

Uman, M. A., *Lightning*, Dover Publications, Inc, New York, 1984.

Uman, M. A., *The Lightning Discharge*, Dover Publications, Inc, New York, 2001.

Amateur Resources

ARRL Technical Information Service sections

Electrical Safety — **www.arrl.org/electrical-safety**

Grounding (various types and topics) — **www.arrl.org/grounding**

Lightning Protection - **www.arrl.org/lightning-protection**

K9YC tutorials on grounding, bonding, RFI, and use of ferrites (**k9yc.com/publish.htm**)

W8JI's web pages on ground systems **(w8ji.com/ground_systems.htm)**

Material Sizes and Values

Table A.1
American Wire Gauge (AWG) Conductor Sizes and Resistance

AWG	*Diameter (inches)*	*Diameter (mm)*	*Area (mm^2)*	*Resistance per 1000 ft (Ω) at 20°C*
0000 (4/0)	0.46	11.684	107	0.049
000 (3/0)	0.4096	10.40384	85	0.0618
00 (2/0)	0.3648	9.26592	67.4	0.078
0 (1/0)	0.3249	8.25246	53.5	0.0983
1	0.2893	7.34822	42.4	0.124
2	0.2576	6.54304	33.6	0.1563
3	0.2294	5.82676	26.7	0.197
4	0.2043	5.18922	21.2	0.2485
5	0.1819	4.62026	16.8	0.3133
6	0.162	4.1148	13.3	0.3951
7	0.1443	3.66522	10.5	0.4982
8	0.1285	3.2639	8.37	0.6282
9	0.1144	2.90576	6.63	0.7921
10	0.1019	2.58826	5.26	0.9989
11	0.0907	2.30378	4.17	1.26
12	0.0808	2.05232	3.31	1.588
13	0.072	1.8288	2.62	2.003
14	0.0641	1.62814	2.08	2.525

Table A.2
Copper and Aluminum Wire Ampacities

Based on ambient temperature of 30 °C (86°F)

Copper Conductors

AWG	*Temperature* 60 °C*	*Rating of* 75 °C**	*Conductor* 90 °C***
14	20	20	25
12	25	25	30
10	30	35	40
8	40	50	55
6	55	65	75
4	70	85	95
3	85	100	110
2	95	115	130
1	110	130	150
1/0	125	150	170
2/0	145	175	195
3/0	165	200	225
4/0	195	230	260

Aluminum Conductors

AWG	*Temperature* 60 °C*	*Rating of* 75 °C**	*Conductor* 90 °C***
14	–	–	–
12	20	20	25
10	25	30	35
8	30	40	45
6	40	50	60
4	55	65	75
3	65	75	85
2	75	90	100
1	85	100	115
1/0	100	120	135
2/0	115	135	150
3/0	130	155	175
4/0	150	180	205

* Types TW and UF
**Types RHW, THW, THWN, THHW, XHHW, and USE
***Types RHH, RHW-2, XHHW, XHHW-2, XHH, THHW, THWN-2, THW-2, THHN, and USE-2

Table A.3
Sheet Metal Thickness (inches)

Gauge	*Mild Steel*	*Aluminum*	*Galvanized Steel*	*Stainless Steel*	*------------ Copper ------------*	
					Thickness	*Weight per sq foot*
10	0.1345	0.1019	0.1382	0.1406		
11	0.1196	0.0907	0.1233	0.125		
12	0.1046	0.0808	0.1084	0.1094		
13	0.0897	0.072	0.0934	0.0937		
14	0.0747	0.0641	0.0785	0.0781		
15	0.0673	0.0571	0.071	0.0703		
16	0.0598	0.0508	0.0635	0.0625	0.0647	48 oz
17	0.0538	0.0453	0.0575	0.0562		
18	0.0478	0.0403	0.0516	0.05		
19	0.0418	0.0359	0.0456	0.0437	0.043	32 oz
20	0.0359	0.032	0.0396	0.0375		
21	0.0329	0.0285	0.0366	0.0344		
22	0.0299	0.0253	0.0336	0.0312	0.027	20 oz
23	0.0269	0.0226	0.0306	0.0281		
24	0.0239	0.0201	0.0276	0.025	0.0216	16 oz
25	0.0209	0.0179	0.0247	0.0219		
26	0.0179	0.0159	0.0217	0.0187		
27	0.0164	0.0142	0.0202	0.0172		
28	0.0149	0.0126	0.0187	0.0156		
29	0.0135	0.0113	0.0172	0.0141		
30	0.012	0.01	0.0157	0.0125		

Wiring for 120 V and 240 V ac Plugs and Outlets

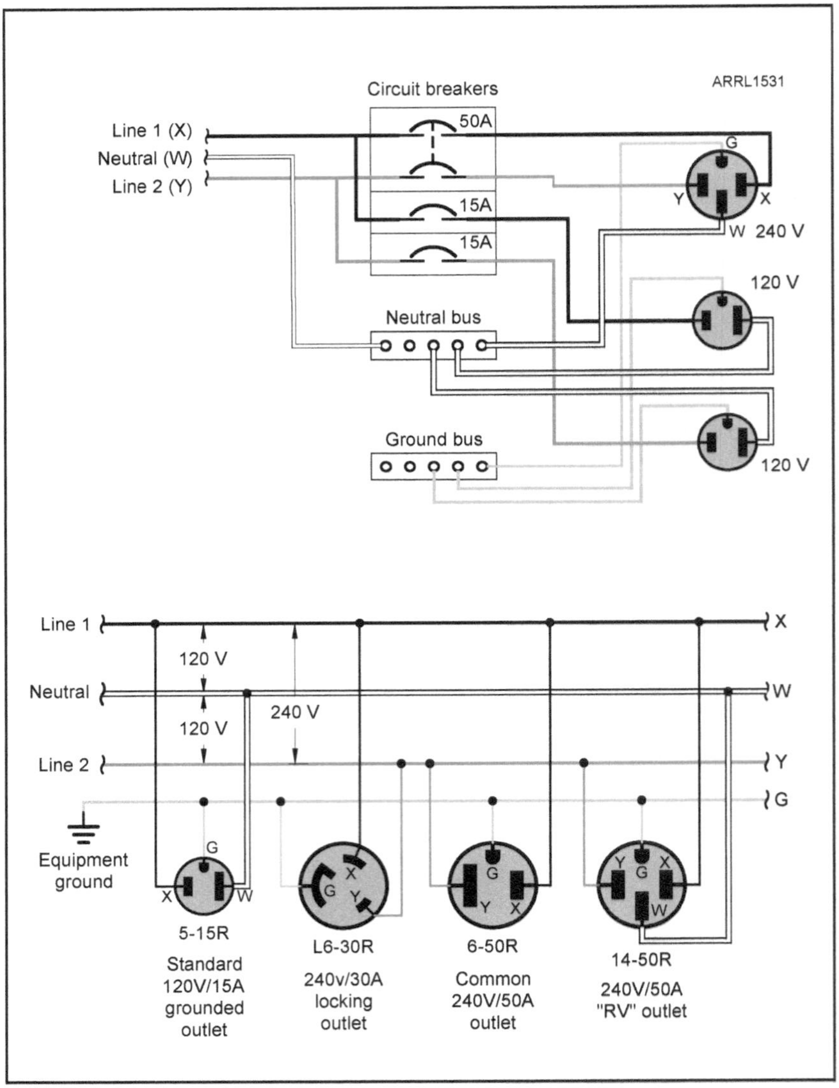

Fig A

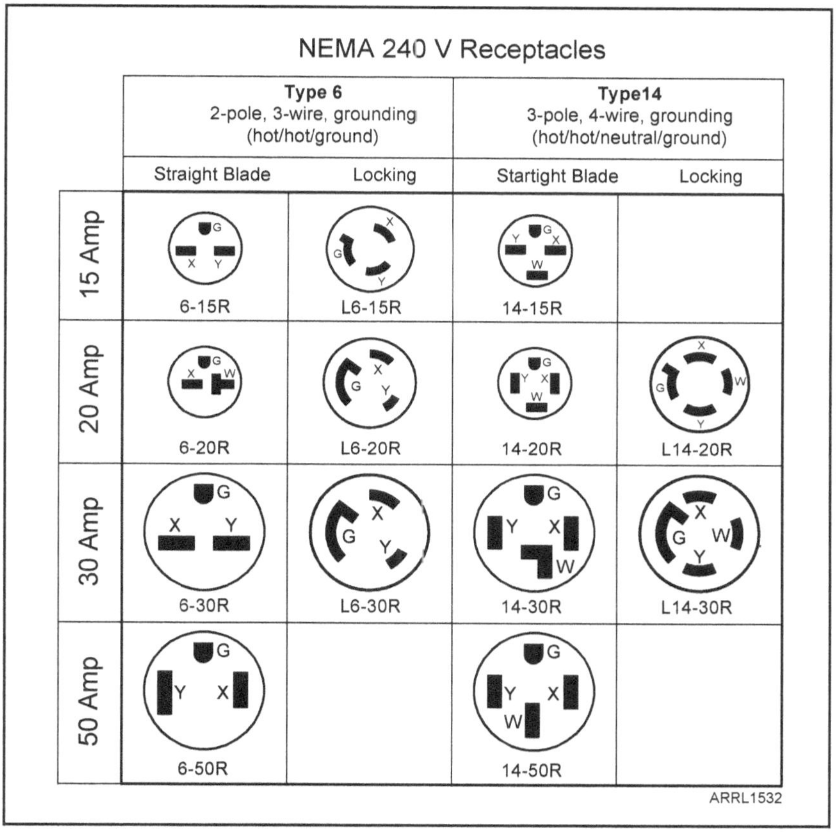
NEMA 240 V Receptacles
Type 6
2-pole, 3-wire, grounding
(hot/hot/ground)
Type14
3-pole, 4-wire, grounding
(hot/hot/neutral/ground)
Straight Blade
Locking
Startight Blade
Locking
15 Amp
20 Amp
30 Amp
50 Amp
6-15R
L6-15R
14-15R
6-20R
L6-20R
14-20R
L14-20R
6-30R
L6-30R
14-30R
L14-30R
6-50R
14-50R
ARRL1532

Fig B

Common-Mode Chokes and Ferrite Characteristics

Table A.4
Ferrite Toroids: A_L Chart (mH per 1000 turns-squared) Enameled Wire

There are differing conventions for referring to the type of ferrite material: #, mix and type are all used. For example, all of the following designate the same ferrite material: #43, Mix 43, 43-Mix, Type 43, and 43-Type.

Fair-Rite Corporation (**www.fair-rite.com**) and Amidon (**www.amidoncorp.com**) ferrite toroids can be cross-referenced as follows:

For Amidon toroids, "FT-XXX-YY" indicates a ferrite toroid, with XXX as the OD in hundredths of an inch and YY the mix. For example, an FT-23-43 core has an OD of 0.23 inch and is made of type 43 material. Additional letters (usually "C") are added to indicate special coatings or different thicknesses.

For Fair-Rite toroids, digits 1 and 2 of the part number indicate product type (59 indicates a part for inductive uses), digits 3 and 4 indicate the material type, digits 5 through 9 indicate core size, and the final digit indicates coating (1 for Paralene and 2 for thermo-set). For example, Fair-Rite part number 5943000101 is equivalent to the Amidon FT-23-43 core.

Ferrite Toroids: A_L Chart (mH per 1000 turns-squared)

Toroid diameter is specified as the outside diameter of the core.

Core Size (in)	*63/67-Mix* $\mu = 40$	*61-Mix* $\mu = 125$	*43-Mix* $\mu = 850$	*77 (72)-Mix* $\mu = 2000$	*J (75)-Mix* $\mu = 5000$
0.23	7.9	24.8	188	396	980
0.37	19.7	55.3	420	884	2196
0.50	22.0	68.0	523	1100	2715
0.82	22.4	73.3	557	1170	NA
1.14	25.4	79.3	603	1270	3170
1.40	45	140	885	2400	5500
2.40	55	170	1075	2950	6850

31-Mix is an EMI suppression material and not recommended for inductive use.

Inductance and Turns Formula

The turns required for a given inductance or inductance for a given number of turns can be calculated from:

$$N = 100\sqrt{\frac{L}{A_L}} \qquad L = A_L\left(\frac{N^2}{10{,}000}\right) \text{ (Amidon cores)}$$

Ferrite Magnetic Properties

$$N = \sqrt{\frac{L}{A_L}} \qquad L = A_L \times N^2 \text{ (Non-Amidon cores)}$$

where N = number of turns; L = desired inductance; A_L = inductance index. Amidon specifies A_L as mH per 1000 turns-squared. See non-Amidon manufacturer's core data sheet to determine appropriate units for L and A_L, usually nH and nH/turns-squared. Make sure you understand which units apply and use the A_L value appropriate for your core.

Property	*Unit*	*63/67-Mix*	*61-Mix*	*43-Mix*	*77 (72)-Mix*	*J (75)-Mix*	*31-Mix*
Initial perm.	(μ_i)	40	125	850	2000	5000	1500
Max. perm.		125	450	3000	6000	8000	Not spec.
Saturation flux density @ 10 oe	gauss	1850	2350	2750	4600	3900	3400
Residual flux density	gauss	750	1200	1200	1150	1250	2500
Curie temp.	°C	450	350	130	200	140	>130
Vol. resistivity	ohm/cm	1×10^8	1×10^8	1×10^5	1×10^2	5×10^2	3×10^3
Resonant circuit frequency	MHz	15-25	0.2-10	0.01-1	0.001-1	0.001-1	*
Specific gravity		4.7	4.7	4.5	4.8	4.8	4.7
Loss factor	$\frac{1}{\mu_i Q}$	110×10^{-6} @25 MHz	32×10^{-6} @2.5 MHz	120×10^{-6} @1 MHz	4.5×10^{-6} @0.1 MHz	15×10^{-6} @0.1 MHz	20×10^{-6} @0.1 MHz
Coercive force	Oe	2.40	1.60	0.30	0.22	0.16	0.35
Temp. Coef. of initial perm.	%/°C (20°-70°)	0.10	0.15	1.0	0.60	0.90	1.6

*31-Mix is an EMI suppression material and not recommended for inductive uses.

Ferrite Toroids—Physical Properties

All physical dimensions in inches.

OD (in)	*ID (in)*	*Height (in)*	A_e	ℓ_e	V_e
0.230	0.120	0.060	0.00330	0.529	0.00174
0.375	0.187	0.125	0.01175	0.846	0.00994
0.500	0.281	0.188	0.02060	1.190	0.02450
0.825	0.520	0.250	0.03810	2.070	0.07890
1.142	0.750	0.295	0.05810	2.920	0.16950
1.400	0.900	0.500	0.12245	3.504	0.42700
2.400	1.400	0.500	0.24490	5.709	1.39080

Different height cores may be available for each core size.
A_e — Effective magnetic cross-sectional area $(in)^2$
ℓ_e — Effective magnetic path length (inches)
V_e — Effective magnetic volume $(in)^3$
To convert from $(in)^2$ to $(cm)^2$, divide by 0.155
To convert from $(in)^3$ to $(cm)^3$, divide by 0.0610
Courtesy of Amidon Assoc. and Fair-Rite Corp.

Table A.5
Transmitting Choke Designs for TFE Coax and Wire-Pair Lines on 2.4-inch OD Type #31 Toroid

(5 kΩ min impedance design)

Freq Band(s) (MHz)	*RG-400 Turns*	*TFE #12 Pair Turns*	*NM/THHN #12 Pair Turns*
1.8	17	17	16
3.5	13	14	13
7	13	13	13
10	12	13	13-14
14	12	12	11
21	11-12 (4.8 kΩ)	11-12 (4.7 kΩ)	11
28	10 (4.4 kΩ)	10 (4.3 kΩ)	10-11 (4.2 kΩ)
1.8-3.5	17	17	16
3.5-10		15	14
3.5-14	13	13	13
7-21	13	12	
High impedance design, if available, given as "Turns (kΩ)"			
1.8	18 (10)	18 (9.5)	18 (9.5)
3.5	16 (8)	15-16 (6.5)	14 (6)
7	14 (6.2)	15 (6.5)	14 (6)
10	14 (6.5)	14 (6)	13-14 (5.5)
14	13 (5.4)	13 (5.5)	12-13 (5)
3.5-14	14 (6, 6, 6, 6)	14 (5.8, 5.8, 5.8, 6, 5)	

Transmitting Choke Designs for TFE Coax and Wire-Pair Lines on 4-inch OD Type #31 Toroid

(5 kΩ min impedance)

Freq Band(s) (MHz)	*RG-400 Turns*	*TFE #12 Pair Turns*	*NM/THHN #12 Pair Turns*
1.8	16	15	15
3.5	13	13	20
7.0	12	15	12-14
10	12	13-14	
14	12		
1.8-3.5	16	21	20
1.8-7	16	15	
1.8-10	16		
3.5-7	19	15	13
3.5-10	14	13	
High impedance design, if available, given as "Turns (kΩ)"			
1.8	23 (17)	22-23 (15)	21-23 (12.5)
3.5	18-20 (11)	16-18 (7.5)	15-16 (6.7)
7	14 (7.5)	13-14 (5.7)	12-14 (5)
1.8-3.5	21 (13,10)	18 (9.5,8)	17 (8.5,6.5)
1.8-7	17 (7, 9.5, 6)	15 (5.5, 7.2, 5)	
1.8-10	16 (5.5, 8.5, 7.5, 5)		
3.5-7	15 (8.5, 7.5)	14 (6.5, 4.8)	14 (6.5, 5)
3.5-10	16 (8.5,7.5,5)	13 (5.8,5.8,5)	

Notes:
Chokes for 1.8, 3.5 and 7 MHz should have closely spaced turns.
Chokes for 14 – 28 MHz should have widely spaced turns.

Table A.6
Coiled-Coax Transmitting Choke Designs

Wind the indicated length of coaxial feed line into a coil (like a coil of rope) and secure with electrical tape.
Lengths are not critical.

Single Band

Freq (MHz)	*RG-213, RG-8*	*RG-58*
3.5	22 ft, 8 turns	20 ft, 6-8 turns
7	22 ft, 10 turns	15 ft, 6 turns
10	12 ft, 10 turns	10 f:, 7 turns
14	10 ft, 4 turns	8 ft, 8 turns
18	9 ft, 6-7 turns	7 ft, 8 turns
21	8 ft, 6-8 turns	6 ft, 8 turns
24	7 ft, 6-8 turns	5 ft, 7 turns
28	6 ft, 6-8 turns	4 ft, 6-8 turns
50*	4-5 turns, 2½" dia	
144,222*	2 turns, 2½" dia	
432*	1 turn, 2½" dia	

Multiple Band (ferrite chokes are recommended)

Freq (MHz)	*RG-8, 58, 59, 8X, 213*
3.5 – 30	10 ft, 7 turns
3.5 – 10	18 ft, 9-10 turns
1.8 – -3.5	40 ft, 20 turns
14 – 30	8 ft, 6-7 turns

*Recommended by GØKSC, **www.g0ksc.co.uk/construction/creatingabalun.html**

Glossary

Ampacity — The current rating of a conductor

Analog ground — In circuits with both analog (linear) and digital circuits, the common reference potential for analog signals

Antenna ground — A system of grounded and other conductors that provides a path to an antenna feed point for RF return current created by a radiated field

Authority Having Jurisdiction (AHJ) — The local government agency having legal authority for establishing building codes and verifying compliance

Battery booster, DC-to-DC booster — A power circuit that supplies dc power at a constant voltage from a battery, regardless of the battery's terminal voltage

Battery Monitoring System (BMS) — A vehicle system that accounts for charging and discharging of the starting battery in order to maintain the battery at full charge

Bonding — Establishing a low-resistance path between equipment and metallic structures to minimize the potential (voltage) between them

Bonding jumper — The conductor used to make a bonding connection, usually heavy and securely connected

Branch circuit — Wiring between the circuit breaker (service entry) panel and loads

Buzz — Undesired low-frequency signals at harmonics of the ac power frequency that are present in an audio or instrumentation circuit

Chassis ground — A point of common reference potential connected to the equipment or assembly enclosure

Circuit ground — A point of common reference potential within an individual circuit. There may be several different circuit grounds within one piece of equipment

Common — A connection point that serves as a voltage reference or current return path

Digital ground — In circuits with both analog (linear) and digital circuits, the common reference potential for digital signals

Electrical service entrance (panel) — The point or enclosure in which a building's ac electrical system is connected to utility power

Electric Metallic Tubing (EMT) — Metal conduit used for electrical wiring

Enclosure — The conductive chassis of a piece of equipment, including exposed hardware, panels, metal case, etc

Engine Idle Shutoff (EIS) — Turning off a vehicle engine when stopped

Equipment ground — See *Safety ground*

Green-wire ground — See *Safety ground*

Ground bus — A heavy conductor connected to a ground system and used as common ground point for connections to multiple pieces of equipment

Ground electrode — A conductor in direct contact with the Earth

Ground loop — A loop formed by grounded conductors, usually refers to a loop in which voltage is induced by magnetic fields from ac currents

Ground (reference) plane — A conducting surface used to maintain a relatively uniform RF voltage

Ground rod — Metallic conductor driven into the Earth to establish a ground connection (see *ground electrode*)

Ground, Earth ground — An electrical connection to the Earth or the conductor making the connection

Grounding — Connecting a piece of equipment to a ground connection

Hum — An undesired sine wave signal in an audio or instrumentation circuit usually caused by pickup of ac power frequency magnetic fields, usually 50/60 Hz

Informational — In a standard, a guideline or suggested practice that is not required (informative)

Lightning dissipation ground — Grounded conductors intended to provide a path to the Earth for charge from lightning

Listed — Devices that are designed and manufactured in accordance with the requirements of a "listing agency," also called a Nationally Recognized Testing Laboratory or NRTL, such as Underwriter Laboratories (UL)

Low-voltage wiring — Circuits carrying ac voltages less than 50 V

NEC — National Electrical Code

NFPA — National Fire Protection Agency

Normative — In a standard, a directive that must be complied with

Overcurrent protection device (OPCD) — Power circuit components such as fuses or circuit breakers that interrupt excessive current flow

Return (power) — The path by which current returns to a power source

RF bonding bus — A heavy conductor used to make electrically-short connections between pieces of equipment

RF ground — An obsolete term used to refer to a connection assumed to be at a constant (zero) RF voltage

Safety Agency (UL, ETL, CSA) — An independent testing body, not affiliated with government, whose business is to test the safety of equipment, fittings, and hardware in their intended use

(AC) Safety ground, Power ground — A connection intended to eliminate a shock hazard by conducting enough current to the system ground that the protective components remove power from the circuit

Signal ground — The reference potential for signal circuits, usually as opposed to power or control circuits

Single-point or star ground — The practice of connecting multiple pieces of equipment to a single point of a ground system

SPGP — single-point ground panel for mounting protective devices and components

Transient — A short impulse of electrical energy

Ufer ground (CEGR) — Ground connection using metallic conductors embedded in concrete. Also known as a concrete-encased grounding electrode

Index

Note: The letters "ff" after a page number indicate coverage of the indexed topic on succeeding pages. Page numbers starting with A (such as A.1) refer to the Appendix.

C

D

E

F

G

H

K

L

R

S

T

Notes

Notes

Notes

Notes

Notes